Favourite Recipes

FROM EUROPE

MARSHALL CAVENDISH

This edition published in 1994 by
Marshall Cavendish Books
(a division of Marshall Cavendish Partworks Limited)
119 Wardour Street
London W1V 3TD

Copyright © Marshall Cavendish 1994
ISBN 1 85435 716 6

Some of this material has previously appeared in the Marshall Cavendish Partwork
WHAT'S COOKING

Printed in the Slovak Republic

C O N T E N T S

Foreword

Favourite Recipes from Europe is a comprehensive introduction to the many cuisines of this area. Practical advice and information on cooking techniques, exciting recipes, serving ideas and background details make it an invaluable reference book for all home cooks.

Favourite Recipes from Europe brings the best recipes from eleven countries. Recipes from distinct areas of the British Isles, France and Italy reflect the contrasts that occur in even one country. Scandinavia has developed independently from the rest of Europe, providing an exciting change to more familiar, Greek and Spanish dishes. For a small country Belgium has a good range of different tastes including those using its famous local beer. Rich Austrian cooking is covered and the distinct flavours from Germany, Portugal and Switzerland are also well represented .

The best traditional meals are reproduced here in individual chapters devoted to a separate nation or region: The Wiener schnitzel of Austria, Scotch broth of the highlands of Scotland and the spaghetti of Italy for example are just a few of the very well known dishes.

With over 150 delicious recipes, each accompanied by a colour photograph and step-by-step instructions, *Favourite Recipes from Europe* will appeal to cookery enthusiasts and gourmets of all levels. Helped and encouraged by the step-by- step photographs and cook's tips, even the most inexperienced cook will be able to create a world-class meal.

Authentic Austria

Creamy cakes and spicy meats make Austrian cooking

rich and distinctive

The true taste of Vienna –Sachertorte with whipped cream (page 12)

*T*HE STRIKING DIFFERENCES between Austria's provinces is reflected in their cooking, though there are many dishes which can be described as typically Austrian rather than purely local.

Viennese delights

Some people maintain that the finest food of all comes from Vienna and certainly no-one would argue that Viennese cakes and pastries are among the finest in the world. The Viennese coffee house has a traditio-nal place in Austrian life and linger-ing deliciously over coffee piled high with whipped cream and a large slice of Sachertorte is a wicked way to spend an afternoon.

Chocolate perils

They say that only in Vienna can the true chocolate taste of Sachertorte be appreciated: Herr Sacher who invented it to please Prince Metter-nich claimed to have flung it together in a moment of exaspera-tion! Chocolate is a favourite ingre-dient – Mohr in Hend for instance is a chocolate and almond pudding masked with hot chocolate sauce.

Provincial pleasures

In the countryside, hearty soups and stews are offered, flavoured with caraway seeds or paprika, allspice and juniper berries. Smoked meats and pickled vegetables recall the days when keeping food was a problem and a typical dish is Bauern-schmaus from the Tyrol – a platter heaped with smoked pork and sausage, sauerkraut, pickles and a large dumpling. Austria is the land of dumplings – in Salzburg liver dumplings are served with brown lentils in winter and consommé in summer and bread dumplings are poached in stews and soups. Würstbraten (see recipe) is served everywhere – the best topside is larded with Frankfurter sausages and served with sour cream gravy.

A specialist shop in the Kärntnerstrasse, Vienna selling Sachertorte

South Tyrolean wine and beef soup

Terlaner Weinsuppe

From the Tyrol, here is a favourite mid-morning 'refresher' for all those engaged in strenuous work, be it felling trees or taking a long walk in the magnificent mountains

- *Preparation: 10 minutes*

- *Cooking: 15 minutes*

500ml/18fl oz beef stock or consommé
5 medium-sized egg yolks
250ml/9fl oz dry or medium-dry white wine
250ml/9fl oz thin cream
ground cinnamon
croûtons fried in butter, to serve

- *Serves 4-6*

- *255cals/1070kjs per serving*

1 Heat the beef stock or consommé to just below boiling point. Meanwhile, whisk the egg yolks together until thick, add the white wine, cream and a pinch of cinnamon in a large bowl.

2 Pour in the heated stock or consommé gradually in a thin stream whisking briskly all the time or the mixture will curdle and separate.

3 Put the soup in the saucepan over gently simmering water or over very low heat, and whisk until the soup is well thickened. Serve the soup in warmed bowls or mugs with croûtons fried in butter and a sprinkling of cinnamon. Fill a thermos for a winter picnic.

SALZBURG SPECIALITY

It would be tempting to think that Mozart who was born in Salzburg enjoyed the local dish of Nockerln, a delicious light concoction of eggs, milk and sugar. It is Napoleon, though, sometimes an exacting gourmet, who is supposed to have caused its invention at a later date, in the course of his Austrian campaign.

Austrian sausage beef

Würstbraten

This Austrian speciality main dish is excellent whether served hot or cold, but omit the sauce if serving the meat cold

● *Preparation: 25 minutes*

● *Cooking: 2 hours*

1.1-1.4kg/2½-3lb joint of topside
4-6 Frankfurter sausages
salt and freshly ground black
  pepper
1 large onion, chopped
2tbls oil
about 15ml/1tbls flour
2tsp paprika
125ml/4fl oz soured cream, to
  serve
Gratin dauphinois potatoes, to
  serve

● *Serves 6-8*

● *390cals/1640kjs per serving*

1 Make 4-6 evenly spaced holes through the joint with the grain of the meat, using a larding needle. Widen the holes as necessary with the handle of a wooden spoon and push the sausages through the holes.

2 Cut off any protruding ends of the sausages and reserve them. Rub the meat with salt and pepper.

3 Sauté the onion in the lard or dripping in a frying pan over medium-high heat. Sprinkle over the flour and stir.

4 Heat the oven to 170C/325F/gas 3. Add the meat to the frying pan and brown the meat on all sides. Transfer the meat and onion to the casserole.

5 Swill out the frying pan with 225ml/8fl oz water and pour this over the meat. Add the paprika and any reserved pieces of the sausages. Cover and put the casserole in the oven for about 2 hours or until the meat is tender, adding a little hot water during cooking if necessary.

6 Transfer the meat to a hot platter and keep it warm. Push the rest of the contents in the casserole through a sieve into a small saucepan, pressing out all the juices. Add the soured cream and allow to thicken over a low heat.

7 Slice the meat so that each slice has rounds of sausage in it. Serve with the cream and dauphinois potatoes.

Cook's tips

The better the meat, the better the finished product. Austrians always insist on the best quality cuts available.

Wiener Schnitzel

The fillet end of leg is the best cut for the escalopes: ask the butcher to flatten it or do it yourself (see Cook's tips, page 12)

● *Preparation: 10 minutes*

● *Cooking: 15 minutes*

4 veal escalopes, 150g/5oz each,
  trimmed and beaten flat
2tbls flour
8tbls fresh breadcrumbs to coat
1 large egg
pinch salt
¼tsp freshly ground white pepper
1tsp olive oil
vegetable oil for frying
25g/1oz butter
1 lemon cut in wedges to garnish

● *Serves 4*

● *395cals/1660kjs per serving*

1 Place the flour and the breadcrumbs on two separate plates. Mix the egg, the salt, pepper and olive oil in a third plate. Put the oil in a large frying pan about 25mm/1in deep, add the butter and place the pan over moderate heat.

2 Dredge the escalopes in the flour, shake off the excess, then dip them in the egg mixture and finally in the

Allow the cake to cool in the tin for
4 5 minutes before turning it out on to
a wire rack and leave to cool for 24 hours.
⏱ The cake is then iced this way up,
bottom uppermost.

5 Warm the apricot jam in a small
saucepan over very low heat. Sieve
the jam and spread it over the sides and
over the flat 'top' of the cake with a brush
or palette knife 6mm/¼in thick.

6 For the icing, melt the chocolate in a
bowl over a saucepan of hot, but not
boiling, water. Remove from the heat.

7 Put the caster sugar in a saucepan
with 4tbls water, bring to the boil,
cook for 1 minute and remove from the
heat. When the sugar syrup is lukewarm,
stir it into the melted chocolate with two
drops of olive oil.

8 Spread the icing while still warm
evenly over the cake using a palette
knife, dipping it in hot water. Allow the
cake to cool before cutting but do not chill
it. This cake is often served piled high with
whipped cream.

COPYCATS AUSTRIAN STYLE

Purists claim that the schnitzel is
only a copy of the Italian veal
escalope dipped in breadcrumbs
rather than grated Parmesan and
insist that Viennese apfelstrudel
was stolen from Hungary – but
it is the Austrian style that
makes these dishes special.

breadcrumbs. Chill for a few minutes.

3 When the fat is very hot, fry the
escalopes, 1-2 at a time, until golden
on one side. Turn them carefully and fry
them on the other side. Drain the meat on
absorbent paper and keep the first ones
warm until all are cooked. Serve them
immediately garnished with wedges of
lemon, parsley sprigs or watercress.

Cook's tips

*To prepare escalopes, make some
incisions around the edges, then flatten
one at a time between sheets of damp
greaseproof paper with a meat bat or
wooden mallet until they are all 6mm/¼in
thick. It is advisable to set the crumbs by
chilling the coated escalopes before frying.*

Sachertorte

The original recipe for this elegant cake
was claimed by both the Hotel Sacher and
Demels, the famous Viennese baker, but
every Austrian housewife has her favourite
version of the recipe. Like many gateaux,
this one is turned bottom uppermost
before icing

- *Preparation: 2 hours*

- *Cooking: 1 hour, plus
  24 hours cooling*

For the chocolate cake:
butter and flour for the cake tin
150g/5½oz plain chocolate
1tbls rum or Madeira
150g/5½oz butter
150g/5½oz icing sugar
6 large eggs, separated
2tsp vanilla sugar
120g/4½oz flour, sifted twice

For the chocolate icing
75g/3oz apricot jam
*120g/4½oz plain chocolate,
  in pieces*
120g/4½oz caster sugar
2 drops refined olive oil

- *Makes 12 slices*   🍴 ££ ⏱

- *400cals/1680kjs per serving*

1 Heat the oven to 180C/350F/gas 4.
Butter and flour 23cm/9in round non-
stick cake tin. Break the chocolate for the
cake into small pieces, add 1tbls water
and place it in a bowl over a saucepan
over low heat. When the chocolate melts,
remove the pan from the heat and stir in
the rum or Madeira until blended.

2 Beat the butter in a mixing bowl until
it is soft and light. Add 115g/4½oz of
the icing sugar and beat until the mixture
is light and fluffy. Gradually stir in the egg
yolks and beat in the barely warm chocolate mixture.

3 Whisk the egg whites until stiff, then
whisk in the remaining sugar and the
vanilla sugar. Fold spoonfuls of the beaten
egg whites into the butter mixture alternately with the flour. Pour the mixture into
the prepared tin. Bake the cake in the
centre of the oven for 1 hour, or until the
cake has evenly risen and springs back if
pressed. Remove from the oven.

Guglhupf

- **Preparation: 1½ hours, including rising time**
- **Cooking: 50 minutes**

110g/4oz butter, softened
100ml/3½fl oz milk
250g/9oz flour, sifted
1 sachet easy blend yeast
2tbls sugar
melted butter and flour for the
    mould
65g/2½oz almonds, blanched and
    slivered
3 large eggs, lightly beaten
65g/2½oz raisins
zest of 1 lemon, grated
1tsp vanilla icing sugar

- **Makes 12 slices**
- **230cals/965kjs per serving**

1 Put the 100g/4oz butter in a bowl. Soften by standing over a bowl or pan

of hot water. Heat the milk in a saucepan over a low heat until just tepid (blood heat).

2 Sift the flour into a warmed bowl with the yeast and sugar. Mix well, then add the milk with the beaten eggs. Mix well to form a smooth batter, gradually adding the softened butter which should be just warm, not hot. Stir in the raisins and the lemon zest.

3 Brush a 900ml/1pt 12fl oz Guglhupf mould with the butter, dust with flour and sprinkle it evenly with almond slivers. Heat the oven to 220C/425F/gas 7.

4 Pour the mixture into the Guglhupf tin (it should be about ⅔ full), cover it with a cloth and stand in a warm place away from draughts to rise.

CROISSANTS FOR BREAKFAST

When the Turks were defeated outside Vienna, the pâtissiers celebrated by baking crescent-shaped rolls. These are said to be the original croissants, though these days they are usually identified more with breakfast in France than in Austria.

5 When the mixture has risen to within 15mm/½in of the top of the mould, place it in the oven and turn down the heat to 190C/375F/gas 5.

6 After 5 minutes, (without opening the door) turn down the oven to 180C/350F/gas 4 and continue baking for up to 40 minutes. Test the cake by pressing the top with your fingertips (it should feel firm) before removing it from the oven. If the top browns too quickly, cover it with buttered greaseproof paper or open the door of the oven slightly during the last few minutes.

7 Allow the cake to cool for 5 minutes before carefully removing it from the mould. Dust it with the vanilla sugar while still warm. This cake is best eaten when fresh. Any left-overs make a delicious trifle.

Walnut gâteau

The Viennese argued as to which was the better cake – *Nusscremetorte* or *Alpenbuttertorte* (nut-cream gâteau or butter gâteau) – until an inspiration suggested the best of both worlds, the basic walnut cake with an *Alpenbuttertorte* filling

- **Preparation: 40 minutes, plus 1 hour drying**
- **Cooking: 50 minutes, plus overnight cooling**

For the walnut cake:
butter and ground walnuts for the
    cake tin
140g/5oz ground walnuts
40g/1½oz fine breadcrumbs
2tbls dark rum
6 large eggs, separated
140g/5oz icing sugar, sifted
For the butter filling:
3 egg yolks
1tsp flour
150ml/5fl oz thick cream
140g/5oz unsalted butter
110g/4oz icing sugar
15g/½oz vanilla sugar
For the icing and decoration:
75g/3oz redcurrant jelly
200g/7oz icing sugar
lemon juice or white rum
12 walnut halves

- **Makes 12 slices**
- **445cals/1870kjs per serving**

1 Heat the oven to 180C/350F/gas 4. Butter a 23cm/9in round cake tin and dust it with ground walnuts. Mix the breadcrumbs with the rum and leave them to soften. Whisk the egg yolks with the icing sugar until they are very thick and creamy. Beat in the breadcrumbs.

2 Whisk the egg whites until stiff but not dry and fold them into the egg yolk mixture alternately with the ground walnuts. Turn the mixture into the cake tin and bake in the oven for 40-50 minutes, until the cake shrinks away from the side of the tin.

3 Allow the cake to cool in the tin for 5 minutes before turning it out on to a wire rack. Leave the cake overnight to cool. 🕐

4 For the filling, whisk the egg yolks, flour and cream in the top of a double boiler over just simmering water (make sure that the water does not touch the top pan) until the mixture is thick. Remove the top pan from the heat and continue whisking until the mixture is cool. Beat the butter with the icing sugar until it is very fluffy, then beat in the egg mixture by teaspoonfuls to make a thick mixture.

5 Slice the cake horizontally into 2-3 layers. Spread the filling between the layers of the cake.

6 For the icing, gently warm the jelly until liquid, then spread it over the top and sides of the cake. Allow the jelly about 1 hour to dry.

7 Beat the icing sugar with enough of the lemon juice or rum to make a stiff paste and spread the paste over the sides and top of the cake. When the icing is just beginning to harden, decorate the cake.

Coffee hedgehogs
Igel

Bought trifle sponges make these a delicious last minute addition to tea-time

● *Preparation: 25 minutes*

8 small sponge cakes
125ml/4fl oz brandy or rum
100g/4oz unsalted butter
30ml/2tbls icing sugar
2tbls strong black coffee
50g/2oz toasted almond slivers

● *Serves 8*   🍴🍴 ££

● *235cals/985kjs per serving*

1 Place the sponge cakes on a serving dish, sprinkle them with brandy or rum and allow to soak and chill in the refrigerator for at least 1 hour.

2 Meanwhile, beat the butter in a mixing bowl until it is soft and light. Add the icing sugar and continue beating until the mixture is light and fluffy. Gradually beat in the coffee, drop by drop.

3 Pile a little of the mixture on top of each sponge cake and coat the sides with the rest. Stick in the toasted almond slivers like hedgehog's prickles.

Variations

The original recipe for icing includes a raw egg yolk which is beaten in with the butter and icing sugar. Before adding raw eggs to recipes consider the consumer.

14

A Banquet from Belgium

The Belgians have a shameless enthusiasm for food, and brew some of the best beers in the world. Not surprisingly, some of their best dishes are a combination of the two

Café liégeois and Strawberry and cheese delight (page 19)

*F*LANDERS, IN THE north of Belgium, includes the provinces of Antwerp and Limburg as well as East and West Flanders. Flemish cooking is based on a rich abundance of excellent raw materials ranging from the superb vegetables around the town of Malines to fish and shellfish fresh from the North Sea and around the coast.

The flavours of Flanders
The Alost region of East Flanders is well known for its excellent onions. The onion soup of Alost is famous outside the region, as is the soft, creamy and slightly sweet onion sauce which goes well with meat.

The North Sea provides West Flanders with a rich abundance of fish, shrimp and mussels. There is a lovely crab soup, croquettes made with crab or shrimps and cod done Flemish-style (see recipe) – simple and delicious. Fillet of sole Blankenberge and Niewpoort herrings are also well known and above all, there are marvellous Belgian mussels which are lovely and plump and a pure delight to eat.

Beer is the national drink of Belgium. There are 355 varieties, with eight types of regional beer made in the town of Antwerp alone, and there are many Belgian recipes which involve cooking fish or meat in beer.

Good food and beer
Brussels and its surrounding province Brabant have always been rich in all the things which make for good cooking – tender chickens, fat calves and prize cattle – plus an abundance of wheat and barley.

Brussels sprouts and chicory (known to the Americans as Belgian endive) find their way into many recipes such as *potage brabançon* and *purée brabançonne*. Chicory is also popular and is eaten raw in salads as well as cooked in a variety of ways. It is wrapped in ham and stuffed with pork and veal; there is braised chicory, creamed chicory, fish with chicory, meat with chicory as well as chicory omelette.

Gourmet's delight

The food of Liège and its province, in the east of Belgium, is substantial and simple but well worth a gourmet's attention for its soups, salads, rich casseroles and stews. Even the sweet combinations are novel, combining strawberries and cheese, or caramel with bread pudding.

If you ask a passer-by in the streets of Liège about the town's most popular dish he will probably answer *'salade'*. The famous *salade liégeoise* (see recipe) is served hot – a delicious mixture of potatoes, cooked French beans, onion and crispy bits of bacon.

Bakers' shops yield a rich variety of specialities, too. The *gâteau de Verviers* is made with yeast, very fine flour, vanilla, sugar, raisins and almonds, while the famous biscuits of Spa contain aniseed, coriander and caraway and are exported all over the world.

Everywhere in Belgium there are countless varieties of waffles and pancakes. Among the most delectable are *bouquettes* which are usually made with buckwheat flour.

A shop scene showing delicious Belgian chocolates

Ardenne vegetable soup

- **Preparation: 35 minutes, plus overnight soaking**
- **Cooking: 3¼ hours**

1 jambonneau or small hock of smoked ham or bacon, weighing about 250g/9oz
1kg/2¼lb green cabbage
1.1L/2pt boiling salted water
100g/4oz streaky bacon, finely chopped
2 carrots, chopped
3 turnips, chopped
1 onion, studded with 2 cloves
1 bouquet garni
salt and pepper
15g/½oz butter

- **Serves 6-8**
- **210cals/880kjs per serving**

1 Soak the jambonneau or hock in enough cold water to cover for 24 hours, replacing the water frequently.

2 Wash and core the cabbage, shred it finely and put it into a large bowl. Pour the boiling salted water over the cabbage, then drain and reserve.

3 Heat the chopped bacon in a large saucepan over low heat until the fat begins to run. Add the reserved cabbage and sauté it in the fat for 1-2 minutes.

4 Add 3L/5½pt cold water to the pan with the hock, vegetables and bouquet garni. Bring to the boil, lower the heat and simmer gently for 3 hours.

5 Remove the hock, the onion and bouquet garni. Add a pinch of pepper, taste and add salt if necessary. Stir in the butter and serve at once.

Cook's tips

If there is any meat on the hock, after simmering remove it from the bone, tear it into pieces and return them to the soup.

COMIC RELIEF

If you visit Liège you will come across a comic figure called *Tchantches* who dominates the town in a rather benevolent way – he has got his own statue, the marionette theatres are always featuring him and his antics, and there's even a soup named after him.

Liège bean and potato salad

- **Preparation: 30 minutes**
- **Cooking: 25 minutes**

450g/1lb French beans, cut into
  bite-sized pieces
salt and pepper
900g/2lb waxy potatoes
25g/1oz butter
250g/9oz streaky bacon, diced
1 small onion, finely chopped
1 sprig fresh tarragon, chopped, or
  ½tsp dried
50ml/2fl oz wine vinegar

- **Serves 4-8**
- **525cals/2205kjs per serving**

1 Cook the beans in boiling, salted water until they are just tender but still crunchy. Drain and keep warm.

2 Meanwhile, boil the potatoes in their skins until just tender. Drain and, when just cool enough to handle, peel and cut them into rounds. Mix the beans with the potatoes in a heated serving dish, cover and keep warm.

3 Meanwhile, melt the butter in a small frying pan over medium-low heat. Add the bacon, frying it gently until crisp.

4 Remove the bacon from the pan with a slotted spoon and add it to the beans and potatoes, along with the onion and tarragon. Pour the bacon fat and melted butter over the salad. Season with salt and pepper to taste.

5 Swill the hot frying pan with the vinegar over low heat, pour it over the salad, blend well and serve immediately.

PURÉED POTATO AND SPROUTS

For a delicious vegetable dish for 4-6 (known as purée brabançonne), purée together 350g/12oz cooked Brussels sprouts, 350g/12oz cooked potatoes and 1 chopped onion softened in butter. Season with salt and pepper and freshly grated nutmeg. Add a knob of butter and a little boiling milk to give the purée a creamy consistency and serve immediately or reheat.

Flemish tomato and egg salad

This beautiful salad speciality of the Convent of Postel in Flanders is excellent to prepare ahead for a starter

- **Preparation: boiling eggs, then 20 minutes**

4 large ripe tomatoes, skinned and
  sliced
3 eggs, hard boiled and sliced
1 shallot or small onion
heart of a small head of celery
1tbls chopped parsley
1 garlic clove, crushed (optional)
3tbls olive oil
1tbls wine vinegar
1tsp Dijon mustard
salt and pepper

- **Serves 4**
- **170cals/715kjs per serving**

200ml/7fl oz strong lager
4tbls dry white breadcrumbs
parsley sprigs, to garnish

● *Serves 4*   ⓘ £

● *345cals/1450kjs per serving*

1 Heat the oven to 180C/350F/gas 4. Butter an ovenproof dish big enough to take the cod fillets in a single layer. Sauté the chopped onions in a large frying pan over medium-low heat in 15g/½oz butter until soft but not browned. Cover the base of the dish with the onions.

2 Dust the fish fillets with the seasoned flour and fry them quickly in 25g/1oz butter in the same pan on both sides to seal.

3 Put the fish on top of the onions. Put a slice of lemon on top of each fillet and dot the fish with the rest of the butter. Add the bay leaf and pour the lager over the fish. Bake for 15 minutes.

4 Sprinkle the fillets with the bread-crumbs and bake for another 5 minutes or until the fish is cooked and the topping is crisp and golden brown. Garnish with parsley and serve at once.

MUSSEL POWER

Belgian mussels are prepared in a great number of different ways ranging from the simple to the very sophisticated. Probably the favourite way of eating them in Flanders is still the simplest of all: they are served raw, like oysters, with lemon or a vinaigrette. To enjoy them in their full splendour, they should be eaten out of doors on a warm evening with a bottle of dry white wine.

◀ *1* Arrange the tomatoes, then the eggs, on a platter.

2 Finely chop the shallots or onion and celery heart and put them in a bowl. Add the parsley and the garlic, if using, and mix well with the oil, vinegar and mustard. Season with salt and pepper to taste.

3 Sprinkle the celery mixture on top of the tomatoes and eggs, cover lightly and chill for 30 minutes or more before serving. 🕐

Cook's tips

The best tomatoes to use for salads are the large, and usually more flavoursome, beef tomatoes. If they are very big you will need only two or three.

Flemish cod with beer

● *Preparation: 15 minutes*

● *Cooking: 30 minutes*

butter, for greasing
4 cod fillets, 175g-225g/6oz-8oz
  each
4 onions, finely chopped
75g/3oz butter
seasoned flour
4 slices of lemon
1 bay leaf

Turkey casseroled in ale

- *Preparation: 10 minutes*

- *Cooking: 2½ hours*

2.3kg/5lb turkey
salt and pepper
150g/5oz butter
1tbls flour
2tbls tomato purée
750ml/1pt 7fl oz pale ale
sprig of thyme or pinch of dried
*   thyme*
2 bay leaves
sprigs of thyme, to garnish
stewed apples and Puréed potato
*   and sprouts, to serve (optional)*

- *Serves 6-8*

- *545cals/2290kjs per serving*

1 Rub the turkey with salt and pepper. Melt 75g/3oz of the butter in a large casserole over medium heat. Slowly brown the turkey on all sides, turning it in the butter. Remove it from the pan.

2 Stir the flour into the butter in the casserole, then add the tomato purée and about 1tbls water. Gradually add the ale, stirring well.

3 Add the thyme, bay leaves and the turkey to the casserole. Add the rest of the butter, cover and cook over very low heat for almost 2 hours, test the legs of the turkey with a skewer. Cook until tender.

4 Transfer the turkey to a warmed serving dish and pour over some of the cooking liquid. Garnish with thyme, then serve with stewed apples and Puréed potato and sprouts, if wished.

BELGIAN BEER

No one visiting Belgium should miss the beers of Brabant, which are not exported to any degree. The best time to sample them is in the summer when they are served accompanied by soft cream cheese garnished with radishes, spring onions and horseradish.

Strawberry and cheese delight

- *Preparation: 20 minutes*
  plus chilling

350g/12oz full-fat soft cheese
2-3tbls icing sugar, plus extra for
*   dusting*
1 egg yolk
225g/8oz strawberries
juice of ½ lemon
sweet biscuits, to serve

- *Serves 4*

- *380cals/1590kjs per serving*

1 In a large bowl, mix the cheese with the icing sugar, add the egg yolk and blend well.

2 Select two-thirds of the best-looking strawberries, put them in a bowl, dust with icing sugar to taste, add the lemon juice and reserve. Press the rest of the strawberries through a nylon sieve.

3 Mix the sieved strawberries with the cheese and put the mixture into a serving dish. Scatter the reserved strawberries on top and pour over any juice that has collected around them. Chill, lightly covered, for at least 2-4 hours. Serve with sweet biscuits such as cigarettes russes.

Variations

Use cottage cheese instead of full-fat soft cheese for a low-calorie alternative. Press it through a sieve to get a very smooth consistency before mixing with the other ingredients.

Café liégeois

One of the best-known Liège specialities, this coffee is in fact just Viennese iced coffee, but was renamed 'liégeois' by the French in tribute to the resistance provided by Liège in the First World War

- *Preparation: 15 minutes*

4tsp sugar
4tbls genever (Belgian or Dutch gin)
strong black hot coffee
4 ice cubes or 4 scoops vanilla ice
*   cream*
150ml/¼pt double cream
icing sugar, sifted

- *Serves 4*

- *190cals/800kjs per serving*

1 Put 1tsp sugar into each of four tall, rather thick, warm glasses. Gently heat 1tbls gin in a metal spoon, light it and pour it over the sugar, repeat.

2 Fill each glass almost to the top with hot, very strong black coffee and add an ice cube or scoop of ice cream to each glass.

3 Sweeten the cream to taste with icing sugar and gently pour it over the back of a large spoon to float on the coffee. Serve the coffee with long spoons.

Cook's tips

Make vanilla sugar simply by storing a vanilla pod in a jar of granulated or caster sugar. But if you haven't any on hand, use extra granulated sugar and add ¼tsp vanilla essence with the egg yolks.

Belgian almond tart

- ● *Preparation: 30 minutes*

- ● *Cooking: 1 hour*

125g/4½oz butter, plus extra for greasing
75g/3oz flour, plus extra for dusting boiling water
125g/4½oz almonds in their skins
125g/4½oz sugar
25g/1oz vanilla sugar (see Cook's tips)
3 eggs, separated
cream, to serve (optional)

- ● *Serves 6*

- ● *455cals/1910kjs per serving*

1 Heat the oven to 180C/350F/gas 4 and butter and flour a 25cm/10in shallow, loose-bottomed flan tin.

2 Pour boiling water over the almonds, leave for 3 minutes, then drain and slip them out of their skins. Spread them on a baking sheet and put in the oven for 15 minutes or until lightly toasted. Leave to cool slightly, then chop finely.

3 Cream the butter with the sugar and vanilla sugar until light and fluffy. Beat in the egg yolks one by one, alternately with the almonds.

4 Whisk the egg whites until they are stiff but not dry, then fold them, a third at a time, into the butter mixture alternately with the flour.

5 Pour into the prepared tin and bake for 40-45 minutes or until golden brown. Turn out and serve warm or cold.

Welcome to Wales

The Welsh kitchen is warm and welcoming,

with freshly made griddle cakes, rabbit stews, rarebits

and, of course, traditional leek dishes

AMONG THE FERTILE green valleys and misty hillsides of Wales, farmhouses can still be found where country cooking has changed little for hundreds of years. Their kitchens are filled with the fragrance of teabreads and drop-scone pancakes, cooked on a heavy iron griddle (called a bakestone or *planc*).

On baking day, a little of the bread dough is reserved and shaped into flat cakes to be cooked on the hot bakestone and eaten warm. This yeasted bread is known as *bara planc*. Sometimes, for a treat, butter, sugar and currants are worked into the dough and it is called *bara brith*, which means literally 'speckled bread'. A speciality from North Wales, it was originally made for special occasions.

Pigs, sheep and leeks

In a country where pigs are second in importance only to sheep, a great deal of lard is available, made by melting down pork belly fat. The melted fat is strained through a fine

Leek broth with bacon (page 26)

sieve into an earthenware jar and used as required for pastry and cake making. Welsh cooking is economical, making the most of its 'home grown' ingredients, such as leeks and lamb. A typical example, perhaps, is the 'miser's feast' (see recipe for Leek broth with bacon).

Perhaps the best-known Welsh speciality is the rarebit (often misspelled 'rabbit'), which is a very tasty snack. A thick melted cheese mixture, including mustard and sometimes beer, is spread on toasted bread and eaten with the fingers. In the olden days, the bread was toasted on a long-handled fork in front of the open fire. Today, it is usually made under a grill, and the toast spread with the rarebit mixture is popped back under the grill to brown. A buck rarebit is served with a poached egg on top and might have a little grated onion added to the cheese topping.

Welsh roots

Root vegetables are extremely popular in Wales. Sometimes they are diced and cooked separately, or mashed together to form a purée, the surface of which is smoothed over, then punched with holes, using the handle of a wooden spoon. The holes are filled with warm cream or melted butter and the dish is called 'punchnep'. It is very delicious if the vegetables are well seasoned with salt, pepper and a little freshly grated nutmeg.

Certain dishes are only made for important occasions. For example, 'shearing cake', flavoured with

Cockle-pickers at work at Penclawdd on the south Gower Peninsula

grated lemon zest and caraway seeds, is made to celebrate the conclusion of sheep-shearing.

TOFFEE EVENING

In parts of North Wales, Toffee Evening used to be a traditional part of Christmas and New Year celebrations. After supper families would make toffee, with everyone helping to 'pull' it.

Mackerel with mussels

- **Preparation: 25 minutes**

- **Cooking: 25 minutes**

4 × 275g/10oz mackerel, cleaned
   and heads removed
salt and pepper
8 streaky bacon rashers
25g/1oz butter, melted
parsley sprigs, to garnish
For the stuffing:
25g/1oz butter
1 onion, chopped
12 mussels, cooked, shelled and
   chopped
2tbls chopped parsley

- **Serves 4**

- **520cals/2185kjs per serving**

1 First, make the stuffing. Melt the butter in a small pan, add the onion and stir well until coated with butter. Cover and cook gently for about 5 minutes or until soft. Stir in the mussels and parsley and season with salt and pepper.

2 Trim off the fins from the fish and season inside with salt and pepper. Divide the stuffing between the fish and re-form them. Wrap two slices of bacon round each mackerel and then place them in a grill pan.

3 Heat the grill to high. Brush the fish with melted butter and grill for 8 minutes on each side, brushing again with butter when turning the fish. Garnish with sprigs of parsley and serve hot.

Lamb in lentil sauce

- **Preparation: 15 minutes**

- **Cooking: 1 hour**

4 large mutton or lamb chump
   chops
3tbls seasoned flour
40g/1½oz bacon fat or butter
2 lean bacon rashers, chopped
450g/1lb swede, diced
300ml/½pt chicken stock
225g/8oz red lentils
salt and pepper

- **Serves 4**   (�) (£)

- **400cals/1680kjs per serving**

1 Coat the chops in the flour. Melt the fat in a large, heavy-bottomed sauce- pan and fry the chops quickly until golden on both sides. Add the bacon to the pan with the diced swede and pour in the stock. Bring to the boil, cover and simmer for 30-40 minutes or until tender.

2 Meanwhile, cook the lentils: remove and discard any discoloured ones, wash well, drain and place in a pan. Cover with cold water, bring to the boil, then simmer for 20 minutes or until soft.

3 Remove the chops, swede and bacon from the pan with a slotted spoon and arrange on a warm serving dish. Keep hot.

4 Drain off any remaining liquid from the lentils and press through a sieve or blend in a food processor. Return to the pan and strain in the liquid from cooking the chops. Stir well and simmer the mixture for 5-10 minutes to reheat. Season with salt and pepper to taste and pour over the chops; serve hot.

WELSH WEDDINGS

'Bidding pie' is a pastry plate pie filled with diced cooked mutton, onion, herbs and seasoning. The pie was traditionally made for a wedding feast; all the bride's friends helped to cook sufficient pies for the whole party. The *gwahoddwr*, or bidder, had the important task of inviting all the guests and recording the amounts of money given by them for pieces of pie. The considerable sums given were the equivalent of wedding presents and were used to furnish the couple's home.

Variations

The Welsh often cook rabbit in this way; substitute four large rabbit portions for the chops and cook for about 1 hour.

FAMOUS SEAWEED

A particularly famous Welsh speciality is laverbread. The laver, an edible seaweed, needs a great deal of washing to remove traces of sand and it is then boiled for up to six hours, until it is quite tender. To make laverbread, it is combined with oatmeal, shaped into round cakes, coated with a little more oatmeal and fried along with the breakfast bacon which gives it extra flavour.

Welsh rarebit

The word 'rarebit' derives from the original 'rare-bite'. The Welsh name for this popular dish is *caws pobi*

- *Preparation: 10 minutes*

- *Cooking: 15 minutes*

225g/8oz Cheddar cheese
50g/2oz butter
15g/¹/₂oz flour
2tsp mustard powder
¹/₂tsp paprika
150ml/¹/₄pt milk
1tsp Worcestershire sauce
salt and pepper
6 large slices of bread
lettuce and flat-leaved parsley,
*   to garnish (optional)*

- *Serves 6*   ↑ ££

- *320cals/1345kjs per serving*

1 Heat the grill to high. Grate the Cheddar cheese. Melt the butter in a saucepan over low heat and add the flour, stirring constantly with a wooden spoon. Add the mustard and paprika and cook for 2 minutes, stirring constantly. Gradually whisk in the milk and cook for 3-4 minutes over low heat, stirring occasionally.

2 Add the Worcestershire sauce and grated cheese. Cook until the cheese has melted, but do not overcook or allow the mixture to boil, otherwise the cheese will become oily. Season with salt and pepper to taste.

3 Toast the slices of bread on both sides. Line the grill pan with a sheet of foil and arrange the toasted bread on it. Spoon the cheese sauce evenly over each bread slice, right to the edges, and grill for a few minutes until the top is bubbling and golden brown. Serve immediately, garnished with lettuce and flat-leaved parsley, if wished.

Stuffed leeks with mustard and cheese sauce

- **Preparation: 15 minutes**

- **Cooking: 1¼ hours**

25g/1oz butter, plus extra for
    greasing
4 large leeks
8 pork chipolata sausages, pricked
25g/1oz fresh white breadcrumbs
For the sauce:
40g/1½oz butter
40g/1½oz flour
425ml/¾pt milk
2tsp French mustard
¼tsp ground nutmeg
175g/6oz Cheddar cheese, grated
salt and pepper

- **Serves 4** ①£

- **575cals/2415kjs per serving**

1 Heat the oven to 200C/400F/gas 6 and grease an ovenproof dish with butter. Trim the leeks so that they are about 18cm/7in long, then split them in half lengthways and wash thoroughly. Drain. Arrange, cut sides upwards and side by side, in the greased dish and dot with the butter. Cover tightly with a lid or foil and cook in the oven for 45-50 minutes.

2 Arrange the sausages in a baking tin and place in the oven with the leeks for 40 minutes or until cooked through, turning once.

3 To make the sauce, melt the butter in a saucepan over low heat. Remove the pan from the heat and stir in the flour. Cook for 1 minute, stirring, then gradually whisk in the milk and bring to the boil, stirring constantly, until the sauce thickens. Simmer for 2 minutes, then blend in the mustard and nutmeg. Stir in three-quarters of the cheese and, as soon as the sauce is smooth again, remove from the heat and season with salt and pepper.

4 Uncover the leeks and place a sausage in the centre of each half. Spoon over the sauce and sprinkle with the remaining cheese and the breadcrumbs. Return to the oven, uncovered, for a further 15-20 minutes or until pale golden. Serve hot.

Variations

Instead of chipolatas, you can use 225g/ 8oz sausagemeat. Working ½tsp marjoram into it makes a tasty addition.

Caerphilly scones

These scones are flavoured with the famous Welsh cheese with a mild, slightly sour taste

- **Preparation: 25 minutes**

- **Cooking: 15 minutes**

50g/2oz butter, plus extra for
    greasing
350g/12oz flour, plus extra for
    dredging
1tbls baking powder
¼tsp salt
100g/4oz Caerphilly cheese, finely
    grated
about 200ml/7fl oz milk

- **Makes 12** ⑪£

- **175cals/735kjs per scone**

1 Heat the oven to 220C/425F/gas 7 and grease a baking sheet. Sift the flour with the baking powder and salt into a bowl and rub in the butter. Stir in the

cheese and, making a well in the centre, add enough milk to make soft dough. Knead lightly.

2 Pat or roll out on a floured surface to a thickness of about 2cm/¾in. Stamp out into 6.5cm/2½in rounds with a plain or fluted cutter.

3 Arrange on the greased baking sheet and bake for about 15 minutes or until well risen and golden brown. Serve warm, split and spread with butter.

Variations

Try using other cheeses such as Cheddar or Cheshire and/or add 1tbls chopped fresh herbs to the mixture; thyme is a particularly good choice of flavouring for this recipe.

Monmouth pudding

- **Preparation: 45 minutes**

- **Cooking: 50 minutes**

275g/10oz day-old white breadcrumbs
300ml/½pt milk
3tbls caster sugar
25g/1oz butter, melted
3 egg whites
½tsp vanilla essence
350g/12oz strawberry or raspberry jam, plus extra for serving (optional)
butter, for greasing
2tbls brown sugar

- **Serves 4**   (♙)(££)

- **560cals/2350kjs per serving**

1 Place the breadcrumbs in a large bowl. Heat the milk to boiling point, pour over the breadcrumbs and leave to stand for 15 minutes.

2 Heat the oven to 130C/250F/gas ½. Lightly fork the sugar and butter into the breadcrumb mixture. In a large, clean, dry bowl, whisk the egg whites with the vanilla essence until the soft peak stage is reached, then fold into the milk and breadcrumb mixture.

3 Spread half the jam in a greased 1.1L/2pt ovenproof dish and cover with half the breadcrumb mixture. Repeat with the other half of the jam and the rest of the breadcrumb mixture, to make another layer. Sprinkle with the brown sugar and bake for 45-50 minutes or until set and golden brown on top. Serve warm, with extra jam if wished.

Leek broth with bacon

This is a complete meal in one, as the bacon and vegetables used to flavour the broth can be served as the next course

- **Preparation: 15 minutes**

- **Cooking: 1¼ hours**

15g/½oz lard
350g/12oz piece of smoked collar bacon
4 large leeks, sliced
1 large potato, diced
1 large carrot, diced
1L/1¾pt well-seasoned lamb or bone stock with all fat removed
1tbls medium oatmeal

- **Serves 4**   (♙)(£)

- **295cals/1240kjs per serving**

1 Melt the lard in a large pan, add the piece of bacon and heat gently, turning, until the fat runs. Add the leeks, potato and carrot, stir well, then cover and cook gently for 5 minutes. Pour in the stock, bring to the boil, cover and simmer for 1 hour.

2 Remove the bacon and vegetables from the pan, place on a warm serving dish and keep hot.

3 Mix the oatmeal to a paste with a little cold water, then add to the broth and bring to the boil, stirring constantly. Simmer for 5 minutes or until thickened. Serve immediately, followed by the bacon and vegetables. Or keep the bacon for another meal.

Hearty Fare from The Shires

Some of England's most solid, satisfying food comes from the Midland 'shires'

Apple batter pudding (page 30)

*T*HIS REGION OF forest, meadow and stream has typified the heart of the English country scene since the days of Robin Hood and his merry men. Although the Midlands are now the industrial centre of England, agricultural traditions continue to flourish and regional dishes are still very popular. From Nottingham and Staffordshire in the north to Gloucestershire in the south and Lincolnshire in the east, the land is fertile and varied, supporting cattle and grain on many different types of soil.

Stews and puddings
The beef cattle of the area provide the meat for many traditional recipes such as the stews and minced beef puddings, as well as the veal used in stuffings and pies.

Regional pies
The Midland Shires are renowned for their cold raised pies, which are filled with either pork, pork and veal or veal and ham. The most famous is the Melton Mowbray pork pie, but almost every town, especially the country ones, has its own slightly individual recipe and shape. Often

fruit, particularly gooseberries and apples, are used as part of the filling, but this shortens the storage life of the pie. Market Harborough pork pie, which contains apple in the filling, is an example of this.

Suet and pastry

Pastry made with suet is often used in both sweet and savoury dishes; this is because so much good beef fat is available for suet. Hot deep dish pies are a popular evening meal, with pastry ranging from simple shortcrust to an elegant buttery puff pastry over a mixture of game with left-over pork products. The humblest pie is made with rabbit or hare when there is any to be had.

Pork 'oddments'

Butchers in the Midlands display many savoury specialities such as faggots, haslet and black pudding, which contain various combinations of pig's fat, left-overs (caul or pluck), blood, oatmeal, onion and potato.

Baker's shop in the Midlands

Market Harborough pie

- **Preparation: 1¼ hours, plus chilling time**

- **Cooking: 3½ hours**

For the hot water crust pastry:
450g/1lb flour
1tsp salt
1 egg yolk
175g/6oz lard
beaten egg, for glazing
For the filling:
1kg/2¼lb minced pork
450g/1lb cooking apples, peeled, cored and chopped
2 onions, finely chopped
50g/2oz caster sugar
½tsp dried sage
1tsp salt
½tsp freshly ground black pepper
For the jellied stock:
1tsp gelatine
150ml/5fl oz strong chicken stock

- **Serves 8**

- **645cals/2710kjs per serving**

1 To make the pastry, sift the flour and salt into a bowl and make a well in the centre. Drop in the egg yolk and cover it with a little flour. Place the lard and 150ml/5fl oz water in a pan and heat gently until the fat has melted. Bring to the boil, then pour quickly into the dry ingredients, mixing with a wooden spoon until the dough is cool enough to handle.

2 Turn the dough out on a lightly floured surface and knead it until smooth and soft and no traces of egg remain. Return it to the warm mixing bowl and cover with a plate. Leave in a warm place for 30 minutes.

3 Meanwhile, mix together the filling ingredients and grease a 15cm/6in loose-bottomed cake tin.

4 Keep the pastry warm during shaping otherwise it becomes hard and difficult to handle. Take three quarters of the pastry and roll it out to a 25cm/10in round. Fold it in half, place it in the tin and mould it carefully and evenly to the shape, raising the sides with your fingers. The pastry should be about 5mm/¼in thick. Leave it to set slightly, then pack in the filling, mounding it very slightly in the centre.

5 Heat the oven to 230C/450F/gas 8. Brush the pastry edges with beaten

egg. Roll out the remaining pastry to make a lid, place it on the pie and seal the edges firmly together. Trim the edges neatly, brush the top with beaten egg and cut a small hole in the centre to allow the steam to escape. Use pastry trimmings to make 'leaves' to decorate the pie. Position them on the pie and brush them with egg.

6 Bake the pie for 20 minutes. Cover it with foil, reduce the oven heat to 170C/325F/gas 3 and bake for a further 3 hours. Cool the pie in the tin.

7 When the pie is almost cold, dissolve the gelatine in the warmed stock and allow it to cool. When the stock is syrupy, pour it into the pie through the hole in the lid, using a small funnel. Refrigerate the pie for 8-12 hours and serve it cold.

Game-keeper's rabbit pie

This recipe tastes equally delicious made with the breasts from 4 plump pigeons instead of the rabbit, or 4 portions cut from the saddle of a young hare

- **Preparation: 45 minutes**

- **Cooking: 1 hour 40 minutes**

4 large rabbit portions (leg if possible)
1 small onion, finely chopped
1tsp chopped parsley
¼tsp grated nutmeg
350ml/12fl oz strong chicken stock
salt and freshly ground black pepper
400g/14oz frozen puff pastry, defrosted
For the forcemeat
20g/¾oz butter or dripping
50g/2oz ham, finely chopped
50g/2oz fresh white breadcrumbs
large pinch of dry mustard
few drops of Worcestershire sauce
1 egg, beaten

- **Serves 6**

- **705cals/2960kjs per serving**

1 Put the rabbit portions in a saucepan with the onion, parsley, nutmeg and stock. Bring to the boil, cover and simmer for about 1 hour, or until the rabbit is tender.

2 Add the salt and pepper to taste to the rabbit mixture and transfer it to a 1L/1¾pt pie dish. Leave it to cool.

3 Meanwhile, make the forcemeat balls. Melt the dripping in a frying-pan over medium heat and fry the ham for 1 minute, stirring. Add the breadcrumbs, mustard, Worcestershire sauce and sufficient egg to give a firm consistency. Reserve the remaining egg for glazing the pastry.

4 Divide the forcemeat into 4 equal portions and shape each with floured hands into a ball. Put the forcemeat balls into the pie dish with the rabbit mixture.

5 Heat the oven to 220C/425F/gas 7. Roll out the pastry at least 4cm/1½in wider and longer than the top of the pie dish. Cut off a strip around the outside of the pastry, 20mm/¾in wide. Dampen this strip and press it to the rim of the pie dish.

ENGLISH CHEESES

Dairy herds prosper on the lush grazing grounds of the Midlands. As a result a variety of excellent cheeses are produced. Amongst these are White Derby, Sage Derby, Leicester and Double Gloucester.

6 Brush the strip on the pie dish with egg, put the lid on and press the edges well to seal. Knock up the edges with a sharp knife and flute them. Brush the top with egg. Decorate with shapes cut from the pastry trimmings, position these and brush with egg.

7 Bake in the oven for 15 minutes then reduce the heat to 190C/375F/gas 5 and bake for about a further 25 minutes, or until well risen and golden brown. Serve hot with green vegetables.

Layered beef suet pudding

(Hough and dough)
The pronunciation of this recipe's name conceals its meaning; hough (pronounced 'huff') means shin of beef, and dough (pronounced 'duff') is the suet pastry

- **Preparation: 35 minutes, plus cooling time**

- **Cooking: 2 hours**

40g/1½oz dripping
2 onions, chopped
450g/1lb minced beef
2 large carrots, grated
2tsp dried parsley
2tbls chopped fresh parsley
25g/1oz flour
500ml/18fl oz beef stock (made with a stock cube if necessary)
salt and freshly ground black pepper
lard or dripping for greasing
For the suet crust pastry
225g/8oz self-raising flour
½tsp salt
75g/3oz shredded suet
50g/2oz raw potato, grated

- **Serves 6**

- **465cals/1955kjs per serving**

1 Melt the dripping in a large pan over low heat, add the onion and fry until soft. Add the beef and fry, stirring, until it looks brown on all sides and crumbly.

2 Add the carrot and both dried and fresh parsley to the pan and cook for a further 2 minutes. Stir in the flour and, when well blended, gradually add the stock and bring to the boil, stirring constantly. Season well with salt and pepper. Leave the mixture to cool slightly while you make the suet crust pastry.

3 Sift the flour and salt into a bowl, stir in the suet and grated potato and

enough cold water to make a firm dough, about 150ml/¼pt.

4 Thoroughly grease a 1.7L/3pt pudding bowl. Divide the suet crust into 4 pieces, weighing about 175g/6oz, 150g/5oz, 100g/4oz and 75g/3oz respectively.

5 Roll out the smallest piece of pastry to fit the bottom of the bowl and spoon in about one third of the meat mixture. Roll out the next largest piece of pastry to fit the bowl and place it over the

BLACK PUDDING

This economical, tasty and nutritious dish has long been popular all over the country. It is made with oatmeal, pork fat and pig's blood. These puddings are often sold in fish and chip shops.

meat. Cover with another one-third of the meat mixture.

6 Roll out the next largest piece of pastry to fit the bowl, place it on the meat and add the rest of the meat mixture. Cover with the last piece of pastry, rolled out to fit. Grease a piece of double-thickness foil, pleat the centre and use to cover the bowl, smoothing the edges well down the sides.

7 Stand the bowl in a pan and add boiling water to come about half-way up the sides of the basin. Cover the pan and keep the water boiling constantly for 1¾ hours, topping up with more boiling water if necessary.

8 Serve hot, turned out, if wished, and cut into wedges, with chopped, lightly cooked cabbage served separately.

PLOUGHMAN'S LUNCH

Typical of this region is the famous ploughman's lunch for farm workers. The original version of this meal consisted of half a freshly baked cottage loaf and a wedge of well matured cheese. This was taken to work and eaten with pickled onions or pickled walnuts washed down with a bottle of ale.

Apple batter pudding

● *Preparation: 20 minutes, plus 10 minutes standing*

● *Cooking: 1 hour*

4 cooking apples
50g/2oz butter
½tsp ground cinnamon
50g/2oz soft brown sugar
butter, for greasing
4tbls golden syrup, warmed, to serve
For the batter:
100g/4oz flour
½tsp freshly grated nutmeg
pinch of salt
3 eggs, separated
350ml/12fl oz milk

● *Serves 4*   🍴 ££

● *500cals/2100kjs per serving*

1 First make the batter. Sift the flour, nutmeg and salt into a bowl and beat in the egg yolks and half the milk. When the batter is smooth, gradually add the remaining milk, whisking all the time. Leave the mixture to stand while you prepare the apples.

2 Heat the oven to 180C/350F/gas 4. Peel and core the apples and drop them into salted water. Mix the butter, cinnamon and sugar together. Butter a 1.7L/3pt ovenproof dish. Drain the apples well, stand them upright in the ovenproof dish and fill the cavities with the spiced butter stuffing mixture.

3 Whip the egg whites until stiff and fold them into the batter. Pour it over the apples and bake in the oven for about 1 hour, or until the batter is well risen and golden brown and the apples are tender. Serve the pudding hot with the golden syrup poured over as a sauce.

Highland Fling

Cross the Cheviots from south to north and you enter a country with strong enduring traditions in law, language, literature – and cooking

Say "Scotland" and people think of porridge, oats, bannocks, haggis, game and shortbread, some of the delights of Scottish cooking, sturdy food with distinctive flavours and strong historical traditions.

"In England, food for horses"

Dr Johnson was really overstating his case when he passed judgment on Scottish oats and there cannot be an Englishman alive who has not fallen ravenously on his "halesome parritch". There was a good reason for the prevalence of oats in Scottish cooking – in general, wheat does not flourish in the cold northern climate whilst oats and barley do well. Hence the oatcake (page 34), the bannock and the barley content of Scotch broth (page 35).

"Wha'll buy my caller herrin'?"

The Scottish coastline is dotted with fishing ports, great and small, and the tasty herring is almost the national dish, rolled in oatmeal and fried or salted and smoked – kippers

Scotch poached salmon with cucumber salad and mayonnaise (page 33)

to the English, but red herrings to the Highlander. Further south, the Tweed and the Tay abound in salmon with a flavour so perfect that plain and simple grilling and poaching are the best ways to cook them (page 33). And let us not forget finnan haddie, invented and perfected at Findon.

Food from France

Scotland had two French queens in

the 16th century, Mary Queen of Scots and her mother, Mary of Guise. With them came French cooks who transformed the plain straightforward cooking of the Scottish court. Hence the delicate veal pie (below); even haggis is supposed to take its name from the French *hâchis*. A leg of lamb is still called a gigot in Scotland and a meat dish or plate is an achete - from *assiette*. As for marmalade - many are those who have tried to make out that it comes from "*Marie est malade*", muttered by Mary Stuart's French chef when he was making orange jam to tempt her appetite. The truth is further away; the word comes from the Portuguese word for quince, *marmelo,* used for jam and jelly from early times.

Scottish sweetmeats

Marmalade lovers much appreciate the bitter, chunky version made in Dundee and Dundee cake is another great favourite, fruit cake flavoured with orange zest and decorated with whole almonds. Black bun is the traditional Scottish New Year cake – rich, dark and laced with brandy. The finest shortbread is said to come from Ayrshire, so rich in butter that it needs no egg to bind it. It is a Scottish speciality which is enjoyed all over the world. So is butterscotch, the delicious hard toffee made with brown sugar and faintly flavoured with lemon juice.

Aberdeen Angus

Scotch beef has been famous since Victorian times – perhaps Queen Victoria's passion for Scotland has something to do with it! The Aberdeen Angus is raised just south of the Highlands and around Aberdeen. It is so deliciously flavoured that it is best served plainly seasoned and grilled. Less expensive cuts are made into mince collops, fresh minced beef combined with oatmeal and pressed into thin cakes before being fried and served with sippets, toasted bread fingers.

The Glorious Twelfth

There is now great competition for restaurants, all over the country, to procure the first of the Scottish grouse on the 12th August, when the season starts.

Hunting on the Scottish moors near Mar lodge

Veal pie

Mary Queen of Scots introduced these delicate flory or Florentine 'tarts' to Edinburgh society. There are sweet versions of these pies too.

● **Preparation: 20 minutes, plus cooling time**

● **Cooking: 40 minutes**

400g/14oz made weight, frozen
    puff pastry defrosted
1 medium-sized egg, beaten
For the filling:
25g/1oz butter
1 onion, chopped
250g/9oz leg of veal, finely chopped
1 cooking apple, peeled, cored and
    chopped
2tbls candied lemon peel, finely
    chopped
150ml/5fl oz chicken stock
pinch of nutmeg
½tsp freshly ground black pepper
1tbls cornflour

● **Serves 4** 🍴 (££)

● **755cals/3175kjs per serving**

1 To make the filling, melt the butter in a small pan over low heat and fry the ▶

onion until soft. Add the veal and stir over moderate heat until it changes colour. Mix in the apple, lemon peel, stock, nutmeg, salt and pepper and bring to the boil. Cover, reduce the heat and simmer for at least 10 minutes.

2 Moisten the cornflour with a little cold water, add to the pan and bring back to the boil, stirring constantly. Simmer for

NEEPS AND TATTIES

Root vegetables grow far better than more tender green vegetables in Scotland's rougher climate and mashed potatoes and swedes – bashed tatties and neeps – are traditional accompaniments to haggis. One green vegetable that grows very well is kale, stronger-tasting than cabbage and often served as kailkenny, mashed up with an equal quantity of potatoes.

2 minutes, remove and cool.

3 Heat the oven to 220C/425F/gas 7. Roll out half the defrosted pastry thinly and use it to line a deep 20cm/8in pie plate. Roll out the remaining pastry to make a lid.

4 Spoon the filling into the pie, brush the edges of the pastry with the beaten egg, put on the lid and seal well all round. Flute the edges and cut a steam vent. Use the pastry trimmings to make pastry leaves or a thistle head for decoration (Master-class, page 37). Brush the pie all over with beaten egg, position the decoration and brush again with egg.

5 Bake in the oven for 10 minutes, then reduce the heat to 190C/375F/gas 5 and continue cooking for a further 20-25 minutes, or until the flory is well risen.

Cook's tips

If veal is difficult or too expensive, use loin or leg of pork.

Poached salmon

The flavour of Scottish salmon from rivers like the Tweed and the Tay is so exquisite that you cannot do better than poach or grill it in the simplest fashion. A tail piece of salmon is favoured because it is often slightly cheaper than a middle cut. Canny Scots tend to poach a large piece of salmon and use up the leftovers in all sorts of delicious ways.

- *Preparation: 5 minutes*
- *Cooking: 30 minutes, plus 2 hours cooling*

1kg/2¼lb tail piece of salmon
2 bay leaves
6 peppercorns

- *Serves 4*
- *455cals/1910kjs per serving*

1 Place the salmon, bay leaves and peppercorns in a fish kettle or large pan and pour over well-salted water to cover. Bring to the boil, cover and simmer for 22 minutes per kg/10 minutes per lb. Allow the fish to become cold in the stock.

2 Lift out the fish and transfer to a board. Remove the skin from the top surface. Serve the salmon in portions with a cucumber salad and mayonnaise.

Cook's tips

A fish stock may be made up by using wine or wine vinegar instead of water. To poach a whole salmon use more water and allow longer to cool.

Venison galantine

Ask the butcher to let you have a small piece of marrow bone and add it to the venison shoulder bone to ensure a good setting stock.

- *Preparation: 45 minutes*
- *Cooking: 3 hours, plus 8 hours chilling*

1kg/2¼lb shoulder of venison on the bone
2 stalks of parsley
1 sprig of thyme
2 black peppercorns
salt and freshly ground black pepper

100g/4oz lean bacon, rind removed
fat or dripping for greasing
225g/8oz minced pork
2 pinches of ground mace
buttered oatcakes, to serve

● **Serves 4**

● **555cals/2330kjs per serving**

1 Remove the venison from the bone with a sharp knife. Put the bones in a pan with the parsley, thyme, peppercorns and 1.5ml/¼tsp salt. Pour over 600ml/1pt water and bring to the boil.

2 Skim, then half-cover the pan and simmer for about 1 hour, or until the liquid has reduced by one-third. Strain the stock carefully and reserve.

3 Cut the venison and bacon into small pieces. Grease a 900ml/1½pt pudding bowl. Put in the venison, bacon and pork in layers, seasoning each layer lightly with salt and pepper and a small pinch of mace. Slowly pour in enough of the stock just to come to the surface of the meat. Cover the bowl with a double thickness of grease-proof paper or foil, tie it firmly with string below the rim.

4 Stand the bowl in a saucepan and pour in enough boiling water to come half-way up the sides of the bowl. Keep the water boiling, cover the pan and cook for 2 hours like a steamed pudding.

5 Lift out the bowl, uncover and if necessary pour in a little more stock to cover the meat. Leave to cool, then press the mixture under a weight ⊙ and chill for about 8 hours. Loosen the galantine, turn it out and serve sliced with buttered oat-cakes and a seasonal salad.

Cook's tips

A galantine of this kind can also be made by substituting veal with bones, turkey or boiling chicken for the venison.

A CUP O' KINDNESS

Whisky, from the Gaelic "water of life", is perhaps Scotland's most famous export and only whisky made in Scotland should really be called Scotch. There are grain whiskies and malt whiskies, straight or blended, made in pot stills or continuous patent stills – Scotch from the western Isles has a distinctive flavour which they say comes from the island peat.

Oatcakes

Bannocks are cut out using a pan lid as a guide. Mark them in quarters or 'farls'.

● **Preparation: 20 minutes**

● **Cooking: 10-15 minutes**

butter or bacon fat for greasing
225g/8oz medium oatmeal
¼tsp salt
pinch of bicarbonate of soda
1tbls melted bacon fat or dripping
about 25ml/4fl oz boiling water
oatmeal to sprinkle

● **Makes about 16**

● **65cals/275kjs per oatcake**

1 Heat a girdle (griddle) or heavy frying-pan and grease it lightly, or heat the oven to 180C/350F/gas 4.

2 Put the oatmeal, salt and bicarbonate of soda in a bowl and stir in the fat and enough water to make a stiff dough. Knead well.

3 Sprinkle a working surface with oat-meal, turn out the dough and roll it out to a thickness of about 3mm/⅛in. Stamp out the dough into 7.5cm/3in rounds.

4 Lift the oatcakes carefully with a slice and cook them on a moderately hot girdle on one side only, until very pale golden underneath and the edges curl ▶

slightly. Alternatively, arrange them on greased baking sheets and bake in the oven for 10 minutes.

5 Handle the oatcakes carefully because they are inclined to crumble. Cool them on a wire rack and when cold store them in an airtight container. If they become soft during storage, toast them lightly or reheat in the oven.

Scotch broth

- *Preparation: 20 minutes*

- *Cooking: 2¼ hours*

700g/1½lb middle neck of mutton or lamb

salt and freshly ground black pepper
2 leeks, chopped
1 large carrot, chopped
1 turnip, chopped
1 onion, chopped
40g/1½oz pearl barley
2tbls finely chopped parsley

- *Serves 6*

- *190cals/800kjs per serving*

1 Cut the meat into neat pieces as far as possible and trim away and discard any gristle and fat. Put the pieces in a large pan and pour over 2.3L/4pt water, season generously with salt and pepper and bring to the boil. Skim the surface, reduce the heat and simmer for 1½ hours.

2 Cut the leeks through lengthways and rinse well. Add the carrot, turnip, onion and barley to the pan. Bring the liquid back to boiling point and simmer for 30 minutes, stirring occasionally. Add the leeks and simmer until the vegetables and barley are soft.

3 Skim any excess fat from the surface of the broth and adjust the seasoning if necessary. Remove the bones if wished. Serve piping hot, sprinkled with the chopped parsley.

Soft and butter buns

- *Preparation: 45 minutes, plus rising*

- *Cooking: baps 15 minutes, 'rowies' 20 minutes*

For the basic dough:
225ml/8fl oz warm milk
1½tsp caster sugar

1 sachet easy blend yeast
about 750g/1½lb strong white flour
1½tsp salt
75g/3oz lard
For the baps:
milk for brushing
flour for sprinkling
For the buttery rowies:
150g/5oz butter
100g/4oz lard

- *Makes 10 baps and 15 buttery 'rowies'*

- *230cals/965kjs per bap*

- *210cals/880kjs per butter bun*

1 Mix together the milk and 200ml/7fl oz warm water. Stir in the sugar and make sure the liquid is blood heat.

2 Sift the flour and the salt into a warm bowl with the yeast and rub in the lard. Make a well in the centre, pour in the liquid and mix to a soft but not sticky dough, add just a little extra flour only if really necessary.

3 Turn the dough on to a floured surface and knead for about 10 minutes, until it is smooth and elastic. Grease a mixing-bowl, return the ball of dough to it and turn once to coat. Cover and allow to double in bulk by standing for a while in a warm atmosphere.

4 Turn the dough out again and this time knead lightly for 2 minutes, until firm.

5 To make the baps, take two-thirds of the bread dough and divide it into 10 equal portions. Roll and shape each into an oval about 5 × 7.5cm/2 × 3in. Arrange on greased baking sheets, allowing room to rise and bake without touching.

6 Cover the baps with greased cling film and allow them to prove in a warm

place until double in size. Heat the oven to 200C/400F/gas 6.

7 Press each bap in the centre with the middle three fingers of one hand then brush with milk and sprinkle with flour. Bake in the oven for about 15 minutes, or until golden brown. The baps should sound hollow when tapped on the base.

8 To make the buttery rowies, take the remaining one-third of bread dough and roll it out to a rectangle 15 × 45cm/6 × 18in with one short side towards you. Beat the butter and lard together then take one third of it and dot over the two-thirds of dough nearest you.

9 Fold down the uncovered third. Then fold up the third nearest you on top. Seal the edges with a rolling pin and give the dough a one quarter turn.

10 Repeat the rolling, dotting, folding, sealing and turning process twice more. Roll out again to a thickness of about 1cm/½in. Cut into 6cm/2½in rounds, or into ovals. Arrange on greased baking sheets, cover with greased cling film and allow to prove in a warm place until double in size. Bake in the oven for about 20 minutes, until golden. Serve warm with sweet or savoury mixtures.

Ginger biscuits

Street traders used to sell these crisp gingerbread biscuits outside the Parliament House in Edinburgh.

● *Preparation: 15 minutes*

● *Cooking: 25-30 minutes*

100g/4oz butter
100g/4oz black treacle or molasses
2tsp ground ginger
100g/4oz soft brown sugar
200g/7oz flour

● *Makes 15*

● *140cals/590kjs per biscuit*

1 Heat the oven to 170C/325F/gas 3. Melt the butter and treacle gently in a medium-sized pan. Remove from the heat and mix in the ginger, sugar and flour until well combined.

2 Press the mixture into a greased Swiss roll tin about 20 × 27.5cm/8 × 11in and bake in the oven for about 20 minutes.

3 Remove the tin from the oven and mark into 15 squares. Leave them to cool until firm, then remove from the tin.

Raspberry and oatmeal cream

● *Preparation: 15 minutes*

● *Cooking: 3-4 minutes*

4tbls medium oatmeal
¼tsp vanilla essence
300ml/10fl oz thick cream, chilled
2tbls caster sugar
225g/8oz raspberries, or canned raspberries, well drained
mint sprigs, to garnish (optional)

● *Serves 2*

● *610cals/2560kjs per serving*

1 Heat the grill to medium-low and when hot, toast the oatmeal lightly on a baking tray. Set the oatmeal aside.

2 Whip the vanilla essence into the cream until it will hold soft peaks, gradually adding the caster sugar.

3 Fold the toasted oatmeal into the cream. Reserve several of the best

berries and fold the rest into the cream.

4 Spoon the mixture into glass sherbert dishes, garnish with the reserved raspberries and mint, if using, and serve.

North Country Know-How

The colder weather of the north of England
with its windswept moors and fells demands hearty food
for large appetites

THIS PART OF the world is a fascinating mixture of beautiful wild countryside, moorland farms, small market towns and great industrial centres dating from the days of the Industrial Revolution. Right across the Pennines from Morecambe Bay to the wild Whitby coast there is a tradition of economical, tasty home cooking that "sticks to the ribs" and keeps you warm in the coldest weather. Stews, hot-pots, pies and pasties are favourites and tasty bits and pieces scorned by southerners are much appreciated here — pork butchers specialise in delicious pork pies of all sizes, pork dripping, scratchings, cold roast pork, chitterlings and haslet and tripe shops sell all kinds of specially prepared tripe. There are fish and chip shops providing the best sup-

Figgy sly cake (page 42)

pers in the country, according to the local opinion.

Yorkshire pudding
They say that only a true Yorkshire-woman can make Yorkshire pudding as it should be made! It is the ideal accompaniment to the Sunday dinner roast beef, though it should

rightly be served separately with meat gravy as a first course. It used to be cooked in the roasting pan under the trivet where the joint sat so that the beef juices could drip into it, but nowadays it is cooked separately and tends to be much crisper – and is usually served on the same plate as the rest of the main course.

Lancashire hot-pot

Basically this is a meat and vegetable stew with a sliced potato topping but there are local variants which make every hot-pot a new experience: Bolton hot-pot (below) has the addition of lambs' kidneys and oysters to make it special. Lakeland hot-pot makes use of the local black pudding, now a delicacy to challenge the *boudin* masters in France. Dumplings are a favourite accompaniment to stews, sometimes flavoured with grated onion or with sage, thyme or parsley.

Bolton hot-pot

- **Preparation: 45 minutes**

- **Cooking: 2 hours**

4 large lamb chops
225g/8oz onion, chopped
700g/1¹/₂lb potatoes, peeled and
    thinly sliced
salt and freshly ground black
    pepper
4 lambs' kidneys
50g/2oz dripping or lard
100g/4oz small button mushrooms
8 smoked oysters
225g/8fl oz stock, made with a cube

- **Serves 4**

- **595cals/2500kjs per serving**

1 Heat the oven to 160C/325F/gas 3. Trim most of the fat from the chops. Put the trimmings in a frying-pan over low heat and cook until the fat runs. Discard the frizzled ends. Turn the heat to medium-high and brown the chops briefly in the fat on both sides.

2 Transfer them to an ovenproof casserole in which they just fit comfortably in one layer. Cover with half the onion and a layer of potato slices and season well. Cover the rest of the potato slices with

Bridlington seafront centres on a still busy harbour with cobles and trawlers.

water to prevent discoloration and reserve.

3 Skin, core and slice the kidneys and add them to the fat remaining in the frying-pan. Toss the slices over high heat for 1 minute, then spread them over the ingredients in the casserole.

4 Melt 15g/¹/₂oz dripping or lard in the frying-pan over medium heat, add the mushrooms and fry for 1 minute, stirring. Arrange them, the oysters and remaining onion over the kidney slices. Sprinkle with seasoning and pour over the stock.

5 Drain and dry the reserved potato slices in a tea-towel and arrange them, overlapping neatly, to cover the top. Melt the remaining dripping or lard and brush it over the potatoes. Cover the casserole and cook in the oven for 1 hour or until potato is tender.

6 Raise the heat to 220C/425F/gas 7 and remove the casserole cover. Cook for a further 30 minutes to allow the potatoes to become golden brown. If making this dish in advance reheat at this stage.

Cook's tips

This recipe was probably made up originally with scrag end or neck chops. If using, allow 30 minutes extra cooking time.

Baked ham in huff paste

Almost every region has its own recipes for curing ham. The famous Yorkshire 'dry' cure was achieved by rubbing the ham thoroughly with coarse salt and sprinkling lightly with saltpetre and Demerara sugar daily for at least 3 weeks. The ham was then washed well in cold water and hung up and smoked for a further 3 weeks. The Yorkshire ham is usually enclosed in a calico bag and stored in a box packed with oak sawdust.

TRADITIONAL LAMB

Not surprisingly, the local wool trade has always meant that lamb and mutton were favourite meats and in days gone by it was the fattier cuts that were preferred. Not only were they cheaper, they provided a simple sauce for floury mashed potatoes and were a useful source of dripping. Nowadays north country housewives prefer their meat well-trimmed.

- *Preparation: 8 hours soaking*

- *Cooking: 3 hours plus 2 hours cooling*

2kg/4¹/₂lb joint smoked ham or gammon
700g/1¹/₂lb flour
about 100g/4oz toasted breadcrumbs

- *Serves 10*

- *740cals/3110kjs per serving*

1 Soak the joint in cold water to cover, overnight, or for a few hours only, if you like it salty. Drain and wipe dry.

2 Heat the oven 200C/400F/gas 6. To make the huff paste mix the flour with about 300ml/¹/₂pt water to form an elastic dough. Roll out the dough to a thickness of 1.5cm/¹/₂in. Dampen the edges of the paste, place the joint in the centre and bring up the sides of the paste to enclose the ham in a neat parcel. Mould the pastry round the ham to cover completely.

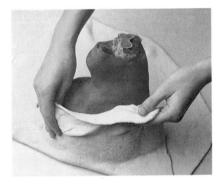

3 Press the edges to seal completely. Put the joint in a greased roasting tin and bake in the oven for 15 minutes. Reduce the temperature to 180C/350F/gas 4 and cook for a further 2¾ hours.

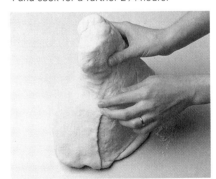

4 Remove the joint from the oven, break off the crust and strip off the rind which sometimes adheres to the crust. ▶

◀ Discard the crust or save for snacks.

5 Cool the ham and place it on the stand ready for carving. Sprinkle the fat evenly with the toasted breadcrumbs. Press lightly to make the crumbs stick to the warm surface.

Cook's tips

For larger joints increase the quantity of huff paste and allow a further 1 hour cooking time for each additional 1kg/2¼lb of meat to be cooked.

Humberside smoked haddock

This East Yorkshire dish is unusual because the smoked fish is accompanied by mashed potatoes. In other regions it is more often poached in milk and served for breakfast with a poached egg on top and accompanied by hot buttered toast.

● *Preparation: making mashed potatoes plus 20 minutes*

● *Cooking: 35 minutes*

450g/1lb smoked haddock, in 4 portions
400ml/14fl oz milk
¼tsp pepper
4 large tomatoes
200g/7 oz flat mushrooms, sliced
50g/2oz butter
25g/1oz flour
700g/1½lb hot mashed potatoes

● *Serves 4*

● *435cals/1825kjs per serving*

1 Heat the oven to 190C/375F/gas 5. Place the fish portions in a shallow oven-proof dish, pour over enough boiling water to cover, allow to stand for 1 minute then drain off. Pour over the milk and sprinkle with the pepper.

2 Pour boiling water over the tomatoes, drain after 10 seconds and skin them, then remove the seeds and juice. Chop the flesh roughly, combine it with the mushrooms and spread the mixture over the fish. Dot the fish with half the butter, cover the dish with greased foil and place in the oven for 25 minutes.

3 Transfer the fish portions to hot plates. Melt the remaining butter in a clean saucepan and stir in the flour. Cook for 1 minute. Add the liquid from the baking dish and stir over moderate heat until the sauce boils and thickens.

4 Place or pipe a mound of mashed potatoes on each plate and make a hollow in the top with the back of a tablespoon. Fill the hollows with some of the sauce and serve the extra sauce separately on the fish.

Apple and bacon plate pie

The pie's simple decoration is an economical use of leftover pastry.

● *Preparation: 20 minutes*

● *Cooking: 1 hour*

225g/8oz flour
pinch of salt
100g/4oz pork dripping or lard
flour for dusting
1 egg yolk
pickled red cabbage, to serve

For the filling:
1 medium-sized cooking apple, peeled
225g/8oz bacon slices, rind removed
1 onion, thinly sliced
100g/4oz potato, sliced wafer thin, then blanched 1 minute
salt and freshly ground black pepper
½tsp dried sage
4tbls light ale

● *Serves 4*

● *720cals/3025kjs per serving*

1 Sift the flour and salt into a bowl and rub in the fat until the mixture resembles breadcrumbs. Add 2–3 tbls water and mix to a firm dough.

2 Divide the pastry in half and roll out one portion on a lightly floured surface to a square large enough to line a 23cm/9in pie plate. Line the plate, trim off the 4 corners and reserve them.

3 Heat the oven to 200C/400F/gas 6. Core and slice the apple, cut the bacon slices into quarters and separate the onion slices into rings. Make a layer each of bacon, onion, apple and potato in the pastry case, sprinkle over salt, pepper and the sage. Spoon over the ale.

4 Roll out the remaining pastry in the same way and use it to make the lid. Beat the egg yolk with 1tsp water and brush it over the pastry edges. Put on the lid and cut a steam vent. Brush the pie with the egg glaze.

5 Decorate the edge of the pie with the reserved 8 pastry triangles, pressing

AMAZING PIES

Famous throughout the north is the Yorkshire Christmas Pye, now no longer made, but it was a great wonder to behold in the olden days.

It was a pie packed with goose and fowl and joints of hare, duck and pigeon. A hot-water crust sealed in the contents and the finished pie was so enormous that it had to be carried on a cart. The largest meat pie in the world was the famous Denby Dale pie, which was made in Yorkshire in September 1988: it weighed 9030 kg.

them carefully into position. Brush with the egg to glaze.

6 Bake the pie for 50 minutes, covering the top lightly with a sheet of foil for the last 10 minutes if necessary, to prevent browning. Serve hot, with pickled red cabbage.

Pikelets

These delicately textured drop scones are usually served for tea. Many recipes exist, some of which include eggs and yeast.

- *Preparation: 10 minutes*

- *Cooking: 10–12 minutes*

250g/9oz plain flour
40g/1¹/₂oz caster sugar
¹/₂tsp salt
¹/₂tsp bicarbonate of soda
about 400ml/14fl oz buttermilk or
*   sour milk*

- *Makes about 18*

- *70cals/295kjs each*

1 Mix together the flour, sugar and salt and make a well in the centre. Dissolve the bicarbonate of soda in a little of the buttermilk and pour into the flour mixture. Beat well, adding enough of the buttermilk to make a thick, smooth batter.

2 Grease a hot griddle or heavy frying-pan and 7.5cm/3in crumpet rings or use a series of plain metal biscuit cutters. Place the rings on the griddle or in the pan and spoon in about 2tbls batter for each pikelet. Cook gently until the bubbles burst and the pikelets are brown. Turn the pikelets and brown the other sides.

3 Transfer the cooked pikelets to a warm plate or ovenproof dish and cover with a clean tea towel while the remaining mixture is cooked into more pikelets. Rub the crumpet rings or moulds with grease before using for each batch of pikelets. Serve hot with butter and jam.

Apple and cheese pie

They say in the North of England 'An apple pie without some cheese is like a kiss without a squeeze'.

- *Preparation: making pastry plus 20 minutes*

- *Cooking: 40 minutes*

400g/14oz made weight shortcrust pastry
500g/18oz cooking apples, peeled, cored and sliced
1tsp cornflour
pinch of ground cloves
75g/3oz caster sugar
75g/3oz Wensleydale or mature Cheddar cheese, crumbled or grated
milk for brushing

- *Serves 6*

- *440cals/1850kjs per serving*

1 Heat the oven to 220C/425F/gas 7. Roll out half the pastry and use it to line a 20cm/8in pie plate.

2 Toss the apple slices in the cornflour, cloves and sugar and place half this mixture on the pastry base. Cover evenly with the cheese and top with the remaining mixture.

3 Roll out the rest of the pastry to make a lid, moisten the edges and seal well together. Flute the pastry edges neatly and cut a steam vent in the top crust. Make decorations from the pastry trimmings if wished. Dampen these, place on the pie and brush all over with milk. Place in the oven for 15 minutes.

4 Reduce the temperature to 180C/350F/gas 4 and continue cooking for a further 25 minutes, or until the pie is golden brown. Serve warm or cold.

Figgy sly cake

- *Preparation: 1 hour plus 4 hours standing time*

- *Cooking: 35 minutes*

225g/8oz dried figs

PIECRUST PROMISES

Sweet and savoury pies are popular in this part of the world and in Lancashire they specialise in flaky pastry filled with moist dried fruit – Eccles cakes which have a rich filling of currants and sugar are well-known all over England but Figgy sly cake (above) is very much a local speciality with its spiced and sweetened pastry crust and rum-soaked figgy filling.

125ml/4fl oz dark rum
40g/1¹/₂oz chopped walnuts
25g/1oz caster sugar
custard or whipped cream, to serve
For the pastry:
350g/12oz flour
¹/₂tsp ground cinnamon
pinch of ground cloves
40g/1¹/₂oz caster sugar
175g/6oz butter
flour for dusting
beaten egg for brushing

- *Serves 8*

- *475cals/1995kjs per serving*

1 Discard any pieces of hard stalk on the figs and chop the fruit. Put them in a bowl and pour the rum over. Stir, then cover and leave to stand for about 4 hours.

2 To make the pastry, sift the flour and spices into a bowl and stir in the sugar. Divide the butter into 4 portions and rub 1 portion into the dry ingredients until the mixture resembles breadcrumbs.

3 Add 4–5 tbls cold water to the bowl, blend with a fork until the mixture is a soft dough. Turn it out on a floured surface and knead lightly until the dough is no longer sticky.

4 Roll out the dough to an oblong about 30 × 15cm/12 × 6in. Dot another quarter of the butter over two-thirds of the strip of dough. Fold over the uncovered third of dough, then fold the opposite third over this as when making puff pastry. Seal the edges with a rolling pin, give the dough a quarter turn and roll it out again to the same size. Repeat the dotting, folding, sealing and turning process twice more, using up the butter. Chill the pastry for 15 minutes.

5 Heat the oven to 220C/425F/gas 7. Cut the pastry in half and roll each piece to a square of about 20cm/8in. Put 1 square on a lightly floured baking sheet.

6 Stir the walnuts and sugar into the soaked figs and spoon this mixture on the pastry on the baking sheet, leaving a clear border all round of about 1.5cm/½in. Brush the exposed pastry border with egg and put the remaining pastry square on top. Seal the edges well and flute them decoratively.

7 Brush the cake with egg and mark the top lightly with the tip of a sharp knife to decorate, but do not pierce the pastry. Bake for about 35 minutes, or until golden brown all over. Serve warm, cut into portions, with custard or whipped cream.

Going Gaelic

Tradition, simplicity and wholesome local ingredients are hallmarks of Irish cooking – ideally combined in the country's famous national dish, Irish stew

*T*HE GRAZING IS lush and extensive in Ireland and the pollution of its rivers, lakes and air is the lowest in Europe. On the heather-covered hills of Kerry and Wicklow counties in the east, sturdy sheep nibble at the young heather shoots which give their meat a fine flavour. The dairy herds in central Ireland have acres of rich grass in which to roam. From their milk comes the famous Irish butter which is used liberally in cooking. What is more, hardly anywhere in Ireland is more than a few miles from a river, lake or sea, all of them teeming with excellent fish and shellfish.

From rivers, lakes and seas

Certain parts of Ireland have their own specialities. The port of Galway is considered the best place for salmon and in the season, the fat salmon can be seen, packed like sardines, lying in the shallows under Salmon weir bridge in Galway town on their way upstream to spawn. Galway is also famous for its oyster beds and at the beginning of the season in September an oyster festival is held. Every pub and hotel for miles around serves oysters, brown soda bread (see recipe), butter and pints of Guinness — Ireland's drink.

The mussels in Ireland, particular-

Herrings potted in Guinness, Leek and oatmeal soup, Soda bread (pages 44-5)

ly at Wexford in county Cork in the south, reach an amazing size and succulence and there are plenty for local eating, although many thousands are exported. This is also the case with scallops, Dublin Bay prawns (now popularly known under the Italian name of scampi) and lobsters in the west. Mackerel are plentiful in the summer months and are often smoked like trout. Herrings, although now in danger of being over-fished, are still available.

43

Irish farmer in mik cart

Classic stew

Another world-famous dish is Irish stew, a dish of mutton, potatoes and onions. It used to be made with young kid, or with the older 'wether', for no farmer would be so foolhardy as to use his young lambs.

The best way to finish a meal – Irish or any other – is with a creamy, whisky-tasting Irish coffee.

Leek and oatmeal soup

● **Preparation: 10 minutes**

● **Cooking: 50 minutes**

6 large leeks, ends trimmed
2tbls butter
1.5L/2½pt milk, or half milk and half chicken stock
2tbls oatmeal flakes
salt and pepper
2tbls chopped parsley
150ml/¼pt single cream, to serve

● **Serves 6**   ① ② ③

● **265cals/1115kjs per serving**

1 Wash the leeks thoroughly to remove the grit. Use most of the green top for this dish, but trim it neatly. Cut the leeks into chunks about 2.5cm/1in long and wash again if necessary.

2 Place the butter and liquid in a large saucepan over medium heat. When the liquid boils, add the oatmeal, boil for 5 minutes, add the chopped leeks and season with salt and pepper to taste. Cover the pan, turn the heat to low and simmer gently for 40 minutes, stirring occasionally. ⏱

3 Add half the parsley and continue cooking for 5 minutes. Serve the soup in warmed bowls with the remaining parsley sprinkled on top and with a little cream poured into each portion.

Soda bread

● **Preparation: 30 minutes, plus standing**

● **Cooking: 40 minutes**

700g/1½lb flour, plus extra for sprinkling
2tsp baking powder
1tsp salt
300ml/½pt buttermilk or sour milk or 300ml/½pt yoghurt and 150ml/¼pt water
1 egg, beaten
oil, for greasing

● **Makes a 900g/2lb loaf** ① ② ③ ④

● **2630cals/11046kjs per serving**

Irish stew and Potato cakes

1 Heat the oven to 190C/375F/gas 5. Mix together the flour, baking powder and salt. Beat the buttermilk, sour milk or yoghurt and water with the egg and stir this into the flour mixture to make a soft but not slack dough. Turn the dough out onto a floured surface and knead for a few minutes until it is quite smooth.

2 Form the dough into a round shape, then cut a cross about 1cm/½in deep into the top of the loaf. Place the loaf on a lightly greased baking sheet and place in the oven for 30-40 minutes. Test the bread with a thin skewer before taking it from the oven to make sure it is done: it should come out clean.

3 Leave the bread for 4-6 hours wrapped in a tea-towel before cutting. ⏱

Cook's tips

The cross on top of the loaf is necessary to allow the raising agent to work – it's not just decorative or a mark of piety!

POTATO PARADISE

Potatoes are one of Ireland's most traditional foods; this fact is reflected in the wide choice of delicious potato dishes. The potato was introduced into Ireland in the 16th century and has since been the staple food. Two famous potato dishes are boxty, a fried potato cake from the north of the country, and colcannon, a mixture of mashed potatoes and kale or, more often, cabbage. Baked potato cakes are a traditional breakfast dish eaten with bacon and egg or sausage.

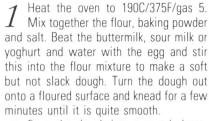

Herrings potted in Guinness

- *Preparation: 15 minutes, plus cooling*
- *Cooking: 50 minutes*

4 fresh herrings, filleted
1 medium-sized onion, sliced
  in rings
1tsp salt
1 bay leaf
4 cloves
6 white peppercorns
6 black peppercorns
1tsp sugar
150ml/¼pt Guinness
150ml/¼pt white malt vinegar
sprigs of parsley, to garnish

- *Serves 4*

- *300cals/1260kjs per serving*

1 Heat the oven to 150C/300F/gas 2. Roll up the fillets from the tail end and arrange them, side by side with the seams underneath, in a casserole. Sprinkle over the onion and salt and add the bay leaf, spices and sugar.

2 Mix the Guinness and vinegar together and pour them over the ingredients. Cover the casserole and place it in the oven for 50 minutes. Leave to cool in the casserole for 2 hours.

3 Transfer to a serving dish, spoon over a little of the liquid and some onion rings and leave to get completely cold. Garnish with parsley sprigs before serving.

Irish stew

- *Preparation: 40 minutes*
- *Cooking: 2 hours*

1.5kg/3¼lb best end neck of lamb
salt and pepper
900g/2lb potatoes, fairly thinly
  sliced
25g/1oz parsley, chopped
1tsp chopped fresh thyme or
  ½tsp dried
450g/1lb onions, fairly thinly
  sliced
butter, for greasing

- *Serves 4-6*

- *690cals/2900kjs per serving*

1 Bone the lamb and place the meat and the bones in a large saucepan,
cover with water and season with salt and pepper to taste. Place the pan over medium heat, bring to the boil, then reduce the heat and simmer for 30 minutes.

2 Drain off the liquid and reserve, allowing it to cool so any fat can be skimmed off the top. While the stock is cooling, cut the meat into fairly large pieces.

3 Heat the oven to 130C/250F/gas ½. Put a thickish layer of potatoes in a casserole, sprinkle with parsley and thyme add a layer of meat, then onions, seasoning each layer well with salt and pepper. Repeat until all the ingredients are used, ending with a layer of potatoes.

4 Pour over the reserved liquid, cover the casserole with greased foil and a lid and place in the oven for 1½ hours or until the potatoes are tender. Add a very little more liquid if the ingredients seem to be drying up during cooking. Serve hot.

Potato cakes

- *Preparation: making the mashed potato, then 30 minutes*
- *Cooking: 20 minutes*

50g/2oz bacon dripping or softened
  butter
250g/9oz self-raising flour, plus
  extra for sprinkling

WEDDING BELLS

On Hallowe'en the Fruity tea bread, or barm brack in the vernacular, is bestowed with oracle powder – a ring is baked in the cake, and whoever gets the slice with the ring will be the first to marry during the year.

200g/7oz potatoes, freshly peeled,
  boiled and mashed
50ml/2fl oz milk
salt
butter, for greasing and serving

- *Makes about 9*

- *235cals/985kjs each*

1 Heat the oven to 200C/400F/gas 6. Mix the bacon dripping or butter with the flour, add the mashed potato and mix well. Add the milk and salt to taste and mix to make a soft, but not slack, dough.

2 On a lightly floured surface, roll out the dough 5mm/¼in thick and cut it into rounds about 7.5cm/3in across.

3 Put the rounds on a greased baking sheet and bake for 15-20 minutes or well risen and golden on top. Serve them hot, split, with butter inside.

Fruity tea bread

- *Preparation: 30 minutes, plus overnight soaking*
- *Cooking: 1½ hours, plus cooling*

450g/1lb sultanas
450g/1lb raisins
450g/1lb soft brown sugar
425ml/¾pt brewed tea
butter, for greasing and serving
450g/1lb flour, plus extra for
  dusting
3 large eggs, beaten
1tbls baking powder
1tbls ground mixed spice
3tbls honey, warmed, to glaze

● *Makes 3 loaves* ① ££ ⊙

● *2045cals/8590kjs per slice*

1 Put the fruit, sugar and tea in a large bowl and soak overnight.

2 Heat the oven to 170C/325F/gas 3. Butter three 1¾pt loaf tins and dust them with flour. Add the flour and beaten eggs alternately to the tea mixture. Add the baking powder and the spice and mix.

3 Divide the batter equally among the prepared tins, level it off with a palette knife and bake for 1½ hours. Test each loaf with a skewer inserted in the centre to make sure it is cooked through; the skewer should come out clean when the bread is cooked..

4 Allow the loaves to cool 5 minutes and turn them out of their tins onto a wire rack. When cold, gently warm the honey, brush it over the tops of the loaves and allow it to set. ⊙ Serve the bread in slices, with plenty of butter.

Irish coffee

Copied all over the world, hot whiskey-laced coffee, drunk through chilled cream, makes a fine ending to a meal as well as a good pick-me-up on a cold morning. Swirl the cream gently over the back of a spoon or it will not float

● *Preparation: making the coffee, then 5 minutes*

2tsp sugar
150ml/¼pt hot strong black coffee
50ml/2fl oz Irish whiskey
1tbls double cream

● *Serves 1* ① ££

● *190cals/800kjs per serving*

1 Warm a 225ml/8fl oz goblet-shaped glass or a large coffee mug by rinsing with hot water. Stir in the sugar, hot coffee and whiskey. Heat a teaspoon in hot water, wipe dry.

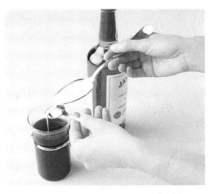

2 Hold the teaspoon, curved side up, across the rim of the glass or coffee mug and pour the cream slowly over the back of the spoon so it floats on the coffee. Drink at once.

BREADLINES

Ireland is unique in that home-made bread is still made with bicarbonate of soda or baking powder. This is because in the uncertain Irish climate, yeast is a less reliable raising agent. The famous soda bread, often called 'brown cake', is still baked daily in countless homes and hotels and even little towns always have bakeries which bake their own bread.

Soda bread is easy to make provided the raising agent is measured properly. For a lighter loaf invert a cake tin over the soda bread, while it is baking in the oven as this helps it to rise.

A Taste of Normandy

Normandy cooks combine simple, fresh ingredients – cream, apples and seafood – with the local cider and apple brandy to produce classic dishes which are among the finest in France

*T*HIS IS A very prosperous part of France, and Norman people are renowned for being hearty eaters. Its position on the north coast is the key to the character of its cuisine. Fish and shellfish fresh from the sea are sold daily at the bustling port of Dieppe, where mussels are a particular speciality. Dieppe gives its name to the classic fish dish Sole dieppoise (see page 49 for recipe): poached fillets of sole in a cream sauce, lavishly garnished with mussels and prawns.

The butter and cream used to make the sauces so characteristic of

Fried Camembert (page 48)

Norman cuisine come from the milk of cows grazing on the lush pastures of the region. Their top-quality milk has earned Normandy the title of France's dairy. Much of it goes to make the world-famous Normandy butter, sweet and unsalted, and a vast range of cheeses, notably Camembert.

Although butter is naturally the main cooking medium, the local cooks also use dripping, known as *graisse normande,* which is made from a mixture of pork and beef fat,

seasoned with vegetables and herbs. This gives a unique taste to meat and poultry as well as pastry.

The meat enjoyed might be superbly flavoured lamb from the sheep which graze on the flat salt marshes beside the sea; chicken from hens who have spent their lives wandering round apple orchards, or game from the country-side (game shooting is a popular sport in Normandy). The region is also renowned for its fine hams, pork pâtés and black and white puddings.

Because Normandy is the most

important apple-growing region in France, apples are always plentiful. Cider is the local drink, almost replacing the wine drunk in the rest of France. When distilled, it becomes the potent apple brandy known as calvados after the district in which it is made. Cider and apples are widely used in the cooking of both meat and fish. For example, sautéed chicken joints are typically served in a cream sauce enriched with cider and apple brandy, finished with fried apple rings.

Apples are also eaten for dessert, often as the filling for a glazed fruit flan, or a sweet omlette, filled with apple and flavoured with calvados.

Fried Camembert

● *Preparation: 15 minutes, plus 1 hour freezing*

● *Cooking: 5 minutes*

8 x 25g/1oz Camembert cheese
   portions
4tbls gooseberry preserve
4tsp lemon juice
2tbls flour
pinch of cayenne pepper
2 eggs, beaten
175g/6oz fresh white breadcrumbs
oil, for deep frying
gooseberry leaves, to garnish

● *Serves 4*   (⁈)(££)

● *475cals/1995kjs per serving*

1 Freeze the Camembert portions for 1 hour. Meanwhile, stir together the gooseberry preserve and lemon juice.

2 Mix the flour and cayenne pepper together. Roll each Camembert portion in the flour, then in the egg and finally in the breadcrumbs.

3 Heat the oil in a deep-fat fryer to 180C/350F or until a 1cm/½in cube of day-old white bread turns golden brown in 60 seconds. Fry the portions of Camembert for 30 seconds or until golden. Drain on absorbent paper and serve at once with the gooseberry preserve, garnished with gooseberry leaves.

Serving ideas

Instead of serving the Camembert as a starter, hand it round at the end of the meal as a savoury. In this case, one portion per person would be ample.

Leeks vinaigrette

● *Preparation: 15 minutes*

● *Cooking: 10 minutes, plus cooling*

8 small leeks, well trimmed
2 parsley sprigs, to garnish
For the vinaigrette:
125ml/4fl oz olive oil
50ml/2fl oz white wine vinegar
1tbls each chopped parsley and
   snipped chives
salt and pepper to taste

● *Serves 4*   (⇡)(£)

● *300cals/1260kjs per serving*

1 Blanch the leeks in boiling salted water for about 10 minutes or until just tender. Drain, then run under cold water until cool enough to handle. Drain thoroughly, and pat dry with absorbent paper.

2 Make the vinaigrette: place all the ingredients in a bowl and whisk with a fork to combine thoroughly.

3 Arrange the leeks side by side in a shallow dish. Pour the vinaigrette over and garnish with parsley sprigs.

Serving ideas

These leeks can be served as a salad; alternatively, accompany them with crusty bread and serve as a starter.

COOKING WITH CIDER

Throughout Normandy, the local cider is used instead of wine to make sauces and marinate meat. It goes particularly well with lamb, heightening its slightly sweet flavour, and is also good with fish, chicken and veal. Cider is also delicious used in sweet dishes, particularly in a syllabub. In fact, try using cider as an alternative in any recipe containing wine, for a subtle variation in flavour — and for economy too.

Cider used for cooking should as a rule be dry, not sweet. Draught cider is ideal, but failing that buy a bottled one and allow it to go flat before using it.

Sole dieppoise

- **Preparation: 1 hour**

- **Cooking: 50 minutes**

450g/1lb small mussels
250ml/9fl oz dry white wine
bouquet garni
4 shallots, very finely chopped
8 × 90g/3½oz sole fillets
1 large fresh thyme sprig
½ fish or chicken stock cube
75g/3oz cooked peeled prawns,
*   defrosted if frozen*
triangular croûtons and finely
*   chopped parsley, to garnish*
For the sauce:
250g/9oz unsalted butter, diced
2tbls double cream
4tsp lemon juice
salt and pepper
pinch of cayenne pepper
pinch of grated nutmeg

- **Serves 4**

- **890cals/3740kjs per serving**

1 Scrub the mussels one by one under cold running water with a hard brush. Pull away the black beards and reserve the mussels in a large bowl of cold water. Discard any that are cracked or have not closed up tightly during cleaning.

2 Bring 75ml/3fl oz wine to the boil in a large, heavy saucepan with a tight-fitting lid. Add the bouquet garni, half the shallots and the mussels and cook for 5 minutes, shaking the pan occasionally. Discard any mussels that are still closed. Remove the cooked mussels from their shells, shaking the liquor back into the pan, and reserve. Discard the shells.

3 Prepare the stock for cooking the sole: (in France this is known as a *court bouillon*): strain the mussel liquor into a measuring jug through a muslin-lined sieve. Reserve 2tbls for the sauce. Add 75ml/3fl oz wine to the jug, then make up to 350ml/12fl oz with water. Pour the liquid into a heavy-based frying pan or

shallow flameproof casserole large enough to take the eight folded fish fillets in a single layer. Add the thyme and ½ stock cube and bring to the boil. Reduce the heat and simmer gently for 5-10 minutes.

4 Start making the sauce: in the top of a double boiler over direct heat, melt

25g/1oz of the butter and fry the remaining shallots until soft but not coloured. Add the remaining wine and reserved mussel liquor and cook until reduced to about 4tbls, stirring frequently. Stir in the cream and carry on cooking until the sauce has reduced to about 3tbls, still stirring. Remove from the heat, cover closely with greaseproof paper to prevent a skin forming, and keep warm over hot water.

5 Cook the fish: fold the fish fillets and gently lay side by side in the simmering *court bouillon*. Poach for 10 minutes, until cooked through but still firm, turning them over with a fish slice halfway through cooking.

6 Meanwhile, gently fry the mussels and prawns in 25g/1oz butter until just hot, then keep warm.

7 Finish the sauce: over simmering water, gradually whisk in the remaining diced butter until smooth and creamy. Remove from the heat and season with lemon juice, salt, pepper, cayenne pepper and nutmeg.

8 Carefully transfer the sole to a heated serving dish. Add the prawns and mussels and spoon the sauce over the dish. Garnish with croûtons and parsley and serve immediately.

Normandy apple tart

- **Preparation: 50 minutes, plus 1½ hours chilling**

- **Cooking: 1 hour 20 minutes**

175g/6oz butter, in very small dice, at room temperature
25g/1oz icing sugar, sifted
1 egg, well beaten
250g/9oz flour
25g/1oz ground almonds
grated zest of ½ lemon
2-4tbls iced water
For the frangipane cream:
2 egg yolks
50g/2oz caster sugar
40g/1½oz flour
250ml/9fl oz milk
25g/1oz ground almonds
1 drop of almond essence
To finish:
350g/12oz tart dessert apples, peeled, cored, thinly sliced and sprinkled with lemon juice
25g/1oz melted butter
1 egg, beaten
50g/2oz apricot jam

- **Serves 6-8**

- **635cals/2660kjs per slice**

1 To make the pastry, blend the butter and sugar with a fork. Stir in the beaten egg. Sift the flour into a mixing bowl and stir in the ground almonds and grated lemon zest. Make a well in the middle and add the butter and 2tbls iced water. Gradually work in the flour and knead to a smooth dough, adding more water if necessary. Wrap well in stretch wrap and chill the pastry for 1 hour.

2 Meanwhile, make the frangipane cream: mix the egg yolks, sugar and flour to a paste in the top pan of a double boiler, away from the heat. Boil the milk, then gradually stir it into the egg mixture. Put the pan onto the double boiler over medium heat, and stir gently until the custard thickens. Remove from the heat, and stir in the almonds and essence. Leave until cold, then chill for 15 minutes.

3 Grease a 23cm/9in loose-bottomed fluted flan tin or a flan ring and baking sheet. Roll out the pastry on a floured board. Line the flan tin with pastry, and prick all over. Trim the edges and chill for 15 minutes.

4 Heat the oven to 200C/400F/gas 6. Cook the pastry case blind: line with greaseproof paper and fill with baking beans. Bake for 15-20 minutes, removing the paper and beans for the last 5 minutes to make sure the base is dry. Leave to cool slightly. Cover evenly with the frangipane cream. Arrange the apple slices in concentric circles on top.

5 Brush the apple slices with melted butter and bake the tart for 35-45 minutes or until the apples are golden brown, covering the pastry edges with foil if necessary. Remove the tart from the oven. Carefully remove the outside ring from the cooked tart and brush the sides and top edges of the pastry with beaten egg. Bake for 4 more minutes, remove from the oven and allow to cool.

6 Melt the jam in a small saucepan with 1tbls water. Brush evenly over the apples and pastry. Leave to cool before serving.

Cook's tips

Make traditional Norman biscuits known as galettes *from the leftover pastry: cut into square or round shapes and bake for 15 minutes below the tart. Dust with icing sugar when cold.*

CHEESES FROM NORMANDY

Normandy produces a host of delicious and varied cheeses. Camembert is the world-famous soft round cheese with a downy white surface and creamy interior. Although Camembert is now made all over France and even in such unlikely places as Japan, the best still comes from Normandy.

Two very ancient French cheeses also come from Normandy: Livarot, similar in shape to Camembert but with a beige crust; and Pont L'Eveque, which is square. Both of these have brine-washed rinds and so are fairly pungent.

Fresh soft cheeses are made all over Normandy: every district has its own with a local name. But because they are perishable few are exported. Exceptions are the factory-made Petit-Suisse and Demi-sel cream cheeses from Gervais. Try eating them French-style, sprinkled with sugar as a dessert.

Ici Paris!

The heart of France is rich in agricultural produce, and Paris,

with its fine restaurants, fine cooking and fine ingredients,

is often heralded as the food capital of the world

Raspberry meringue ring (page 56)

*T*HE RICH AGRICULTURAL region, of which Paris is the heart, is known as the Ile-de-France and is the historic centre of the country. The capital's 'food basket', it is dominated both by the city and by its suburbs.

Its presence, plus the nearness of great châteaux like Versailles and Fontainebleau to the south and Chantilly (of whipped cream fame) to the north, has set the most demanding standards in food which this generous region has been able to meet.

Village fare
Food may now come from further afield, but at one time the capital was dependent on its nearby villages for its fresh produce. Montreuil was known for its peaches, Montmorency for cherries, Vaugirard for strawberries, St Germain and Clamart for peas, artichokes and butter. Although these villages have now been absorbed by the city's ever-growing suburbs, their names are immortalized in classic recipes.

Dairy centre
In the Ile-de-France, butter and cream are added to dishes with a lavish hand. Part of Brie, where France's best-loved cheese is produced, is also in the region, for the Ile-de-France has been a dairy as

well as an agricultural centre for the capital.

Fontainebleau also has its own cheese, Crémet de Fontainebleau. Perhaps less famous, it is nonetheless creamy, slightly sweet and delicious.

To the south and east is the province of Orléanais, with its wooded countryside and many ponds and lakes. It is bordered by the Loire and some of the best shooting in France takes place here. The area is famous for its game pâtés, that of Chartres being the most notable. The middle Loire is prized for its fish, such as pike, while crayfish are also found locally.

Gourmets' paradise

Many of these influences have found their way to Paris which, because of its large population and wealth, has always drawn chefs, pâtissiers and other employees of the food trade to its doors. Their regional culinary traditions have mixed with those of the city itself to offer such variety and largesse that every gourmet dreams of visiting the capital.

A luxurious display at Fauchon, an exclusive food shop in Paris

HOW IT ALL BEGAN

The word 'restaurant' is over 200 years old and was devised when a Parisian innkeeper began selling bowls of bouillon as a 'restauratif'.

Montmartre onion soup

- **Preparation: 25 minutes**

- **Cooking: 40 minutes**

2tbls oil
350g/12oz onions, thinly sliced
1tbls flour
about 150ml/¼pt dry white wine
salt and pepper
3 egg yolks
1tbls double cream
100g/4oz toasted croûtons
grated Gruyère cheese, to serve

- **Serves 4** （♈)(££)

- **430cals/1805kjs per serving**

1 Heat the oil in a large saucepan, add the onions and fry gently, stirring occasionally, until golden. Sprinkle in the flour, stir well and cook for 1 minute.

2 Add the wine and 1L/1¾pt water, season with salt and pepper and bring to the boil, stirring constantly. Cover and simmer for 30 minutes.

3 Beat together the egg yolks and the cream. Remove the soup from the heat, beat 4tbls into the egg mixture and pour this back into the saucepan, whisking constantly off the heat.

4 Put the croûtons into a warmed tureen, pour in the soup and serve at once, with the grated cheese.

Brie cheese puffs

- *Preparation: 20 minutes*

- *Cooking: 20 minutes*

225g/8oz Brie
1 egg yolk
2tbls double cream
15g/¹/₂oz butter
20g/³/₄oz flour
6tbls milk
1tsp Dijon mustard
salt and pepper
375g/13oz puff pastry, defrosted if
*   frozen*
1 egg yolk, beaten

- *Makes 9*   (**) (££)

- *305cals/1280kjs per puff*

1 Heat the oven to 220C/425F/gas 7. Trim off and discard the rind from the Brie, then mash the cheese. Stir the yolk into the cream.

2 Melt the butter in a small pan over low heat and stir in the flour. Cook for 30 seconds, stirring. Gradually add the milk and cook, stirring all the time, until the sauce thickens. Season with mustard and salt and pepper to taste, and mix in the cheese. Stir until melted, then remove from the heat, add the egg yolk and cream and beat well. Set aside and allow the mixture to cool.

3 On a lightly floured surface, roll out the pastry thinly and cut into nine equal squares. Divide the cheese filling among the pastry squares, placing it on one half of each piece.

4 Dampen the edges of the pastry, fold over and press the edges firmly together to make triangles. Arrange the pastries on a damped baking sheet and brush with beaten egg yolk to glaze. Bake for about 15 minutes or until golden brown. Serve hot.

Variations

Coulommiers can be used instead of the Brie. A member of the same family, it is a soft paste cheese which looks like an overgrown Camembert.

READING THE MENU

It can sometimes be puzzling understanding recipe titles on a menu. This is because some of the words have no direct translation, being instead place names associated with particular produce. Crécy, for example, signifies carrots as in *oeufs à la Crécy*, which combines eggs with a carrot purée; St Germain indicates peas, as in *potage St Germain*, a soup made with fresh or dried peas; Clamart also denotes peas, as in *oeufs Clamart:* peas and scrambled eggs served with toast.

Veal rolls in mushroom sauce

- *Preparation: making duxelles,*
  then 35 minutes

- *Cooking: 40 minutes*

4 large or 8 small veal escalopes
*   (about 450g/1lb total weight)*
1tbls chopped onion
1 small shallot, chopped
100g/4oz minced pork
salt and pepper
25g/1oz melted butter, plus extra
*   for greasing*
300ml/¹/₂pt beef stock
1tbls grated Parmesan cheese
flat-leaved parsley, to garnish
For the sauce:
40g/1¹/₂oz butter
25g/1oz flour
duxelles (see Cook's tips, page 54)
4tbls double cream

- *Serves 4*   (**) (££)

- *375cals/1575kjs per serving*

1 Heat the oven to 190C/375F/gas 5. Beat out the escalopes until they are almost twice their original size. If you are using large ones, divide them in half lengthways. Trim them neatly and finely chop the trimmings.

2 Mix together the onion, shallot, veal trimmings, pork and salt and pepper to taste. Divide the stuffing mixture among the escalopes, placing it at one end of each escalope. Roll up neatly and tie each veal bundle with fine white thread.

1 Peel and cut the potatoes along the grain into 3mm/⅛in slices. Cut each slice into a rectangle approximately 7.5cm/3in × 4cm/1½in, then cut off the corners to make a lozenge shape.

2 Heat the vegetable oil to 140C/275F. Drop the slices into the oil, about ten at a time. They will sink, then rise. As they are rising, shake the pan.

Veal rolls in mushroom sauce
(page 53)

◀ 3 Place the veal rolls in a greased shallow ovenproof dish, brush with melted butter, cover and cook in the oven for 15 minutes. Spoon over the stock, cover again and return to the oven for a further 15 minutes. Strain off the juices and reserve. Keep the veal rolls hot in the covered dish.

4 To make the sauce, melt the butter and stir in the flour. Cook for 1 minute, stirring all the time. Gradually add the reserved juices and bring to the boil, stirring constantly. Mix in the duxelles and cook gently for 3 minutes, stirring all the time. Remove from the heat, blend in the cream and season. Heat the grill to medium.

5 Transfer half the sauce to a shallow heatproof dish and put the veal rolls on top. Spoon the remaining sauce over and sprinkle with the cheese. Put the dish under the grill until golden brown on top. Garnish and serve very hot.

Cook's tips

To make duxelles, melt 15g/½oz butter in a saucepan, add ½ chopped onion or 1 chopped shallot and cook over low heat for 10 minutes until soft. Add 175g/6oz chopped mushrooms and stalks and cook for about 15 minutes, stirring constantly, until all the liquid has evaporated. Add 1½tbls finely chopped parsley and cook for 2 minutes, or until very dry.

Souffléd potatoes

This is a tricky recipe; it's very important to get the oil temperature right and to fry the potatoes the second time almost immediately after the first. Even the most experienced cook should expect some of the potato slices to be failures – but don't panic; small portions will suffice

- **Preparation: 15 minutes**

- **Cooking: 20 minutes**

700g/1½lb potatoes
vegetable oil, for deep frying
salt

- **Serves 4**

- **375cals/1575kjs per serving**

3 Cook for 4-6 minutes, until the potatoes are turning clear in the centres, show a difference in texture around the edges and start to blister.

4 Drain the potatoes on absorbent paper, then leave them to cool for 5 minutes. Reheat the oil to 196C/385F, then drop the potato slices, about ten at a time, into the oil. The slices should puff up at once. Fry until golden, drain on absorbent paper, sprinkle with salt and serve.

Duck with spring vegetable sauce

- **Preparation: 50 minutes**

- **Cooking: 1 hour**

1.9kg/4¼lb duckling
salt and pepper
1 tbls brandy
75g/3oz lean minced pork
75g/3oz minced veal
1 carrot, chopped
1 onion, chopped
2 tbls chopped parsley
75ml/3fl oz dry white wine
juice of ½ lemon
75ml/3fl oz veal or chicken stock
2 hot cooked or canned artichoke
  bottoms, 1 lemon, quartered, and
  flat-leaved parsley, to garnish

- **Serves 2**

- **590cals/2480kjs per serving**

1 Heat the oven to 190C/375F/gas 5. Divide the duck into four portions; reserve the two breast portions.

2 With a very sharp knife, make a cut lengthways through the flesh of each leg (still attached to the rest of the leg portion) to the bone. Carefully cut away the flesh, with the top of the knife always next to the bone so you can ease the bone and cartilage out of the leg portion.

3 Sprinkle the boned leg portions with salt, pepper and the brandy. Mix together the minced pork and veal and season well with salt and pepper. Stuff the boned legs with the mixture, then sew them up securely with a needle and thread or secure with several small skewers.

4 Put the carrot, onion and parsley in an ovenproof dish, lay the stuffed duck portions on top, cover and cook in the oven for 45-60 minutes.

5 Meanwhile, carefully cut away the bone from the duck breasts and skin them. Poach the breast fillets in the wine, lemon juice and stock, covered, for 5 minutes or until they are cooked but still pink. If you prefer the flesh to be less pink, continue cooking for a further 5 minutes. (Use the duck trimmings for another dish.)

6 Drain the breast fillets, reserving the stock, and place them on a serving dish with the stuffed legs. Keep hot.

7 Reduce the stock from cooking the fillets by half, add the vegetables and juices from the ovenproof dish and cook for 2 minutes. Skim off the surplus fat and discard it, then press the mixture through a sieve or purée it in a blender.

8 Taste and adjust the seasoning. Remove the skewers or thread from the leg portions. Spoon the sauce over the duck portions and serve garnished with the artichoke hearts, lemon wedges and flat-leaved parsley. Serve very hot.

POMMES PARISIENNE

A popular side dish is Parisian potato balls, made by using a large melon baller to scoop out rounds of potato. These are rinsed, drained, par-boiled for 1 minute, then rinsed and drained again. To complete their cooking, they are sautéed in butter, the pan being frequently shaken and the potatoes carefully turned with a slice. The potatoes are ready when they are golden brown and cooked through to the centre.

Traditionally, the potato trimmings are boiled in salted water until tender, seasoned, mashed with hot milk and then piped as a border garnish.

Stuffed pigeons

- **Preparation: making the Béchamel, then 50 minutes**

- **Cooking: 1 hour 30 minutes**

4 small, plump pigeons
125ml/4fl oz Béchamel sauce
   (make in same way as in step 7)
175g/6oz lean minced veal or pork
50g/2oz chicken livers, chopped
150g/5oz butter
salt and pepper
350g/12oz shelled peas
1 small lettuce heart, quartered
1tsp sugar
225g/8oz button mushrooms
40g/1½oz flour
225ml/8fl oz beef stock
125ml/4fl oz Madeira

- **Serves 4**

- **835cals/3505kjs per serving**

1 To partially bone the pigeons, split each one along the back and open out. Cut out the backbone and ribcage, cutting against the bones carefully with a very sharp, thin knife, but leave in the leg and wing bones to retain the bird's shape.

2 Heat the oven to 180C/350F/gas 4. Mix the Béchamel sauce, veal or pork, chicken livers and salt and pepper to taste.

3 Divide this stuffing mixture among the pigeons, re-shape them and sew up the openings, secure with skewers or tie into shape with fine white string.

4 Melt 40g/1½oz of the butter in a deep flameproof casserole over medium-low heat, put in the birds and brown them slowly on all sides, turning frequently. Season them lightly with salt and pepper, cover and put in the oven for 1¼ hours.

5 Meanwhile, cook the vegetables. Melt 25g/1oz of the butter in a saucepan over low heat, add the peas, lettuce, sugar and 4tbls water. Stir, season lightly with salt and pepper, cover and cook gently for 15 minutes over very low heat, shaking the pan occasionally but not lifting the lid.

6 Melt 25g/1oz butter in a saucepan over medium heat and lightly sauté the mushrooms in it until pale golden.

7 To make the sauce, melt the remaining butter in a small saucepan and allow it to become golden brown before stirring in the flour. Cook gently, stirring all the time, until brown, but do not let it get too dark. Gradually blend in 175ml/6fl oz stock, stirring constantly, until thickened.

8 Add the Madeira, stirring briskly all the time. Keep the sauce barely simmering for 3 minutes and add salt and pepper, if necessary. Keep the sauce hot.

9 Combine the pea mixture and mushrooms, pile up in the centre of a warm serving platter and keep warm. Cut the pigeons in half with a very sharp knife, removing any string or skewers, and arrange round the vegetables.

10 Pour the remaining stock into the casserole and boil hard for 1 minute, scraping up any sediment. Spoon the liquid over the birds. Transfer the sauce to a warm sauceboat to serve.

Raspberry meringue ring

This dish, known as *Meringue royale*, is a culinary joke at the expense of Marie-Antoinette. Her fondness for meringues was well known. The ring is decorated with a raspberry 'necklace' in an allusion to jewels that disappeared and, in doing so, caused a scandal

- **Preparation: 40 minutes**

- **Cooking: 1 hour**

3 egg whites
175g/6oz caster sugar
For the filling and decoration:
225ml/8fl oz double cream
1tsp caster sugar
2-3 drops of vanilla essence
350g/12oz large raspberries
500ml/18fl oz vanilla ice cream

- **Serves 6**

- **415cals/1745kjs per serving**

1 Heat the oven to 140C/275F/gas 1 and line two baking sheets with non-stick paper. Draw a 20cm/8in circle on each piece of paper.

2 Put the egg whites in a large bowl and whisk until stiff. Start adding the sugar, 1tbls at a time, whisking vigorously after each addition until the meringue stands in firm glossy peaks.

3 Pipe or spread about two-thirds of the meringue into one of the marked circles to make a round. Fit a piping bag with a plain 2cm/¾in nozzle and spoon in the remaining meringue. Pipe meringue just inside the marked circle on the second sheet of paper to make a ring.

4 Bake the meringues for 1 hour or until they are crisp and dry. Leave to cool on a wire rack, then remove the paper from the meringue bases.

5 To make the filling, whip the cream until soft peaks form. Add the sugar and vanilla essence and continue whisking until the cream holds its shape.

6 Put the large meringue on a serving plate and spread over two-thirds of the cream. Put the meringue ring on top. Fill the centre with half the raspberries and cover with scoops of ice cream. Put the remaining cream in a piping bag fitted with a star nozzle and pipe rosettes all round the ring of meringue. Top each rosette with a raspberry, leaving a gap at one point so that the raspberries look like a necklace. Serve immediately.

Cuisine Provençale

With its staples of tomatoes, anchovies, olives and wild herbs, provençal cooking conjures up images of fishing ports, olive groves, lavender fields and a deep blue Mediterranean sky

*O*F THE FRENCH regions bordering on the Mediterranean, Provence is most like the travel brochure's image of the south of France. It is a land of fiercely sunlit fields and of pine-dotted slopes where vineyards, olive and lemon groves flourish. From walled hilltop villages bright with pink and purple bougainvillaea, one can catch glimpses of the Mediterranean lapping the coastline.

The origins of provençal cookery go back to the ancient Greeks and Romans who established the olives and vines over the hills and made use of the aromatic wild herbs to flavour their fish and meat. These traditions were later adapted and modified by the French, who in turn introduced them to the north of the country. Ever since, provençal cooking has been a major influence on all French cookery.

Italian heritage

Nice and its neighbouring coast, for instance, came under Italian rule during the Middle Ages and niçoise cooking reveals more than a hint of the Italian past: it's based on olive oil and relies for flavour on the pun-

Salade niçoise (page 61)

gent, sun-dried herbs that grow wild.

Salade niçoise and *Ratatouille* (see recipes) are both internationally known classics. But the locals still stay faithful to many of their own highly original ways of combining and preparing food, like mixing deep-fried sardines with olives or stewing veal with anchovy fillets.

Other typical niçoise dishes are *pissala*, pickled whitebait; *stoccaficada*, dried cod cooked with tomatoes and potatoes; *lièvre à la niçoise*, hare stewed with sausages, mushrooms and onions; and *pissaladière*, onion flan with anchovy fillets (see recipe). Desserts are mostly made from local fruits: *fougassettes*, sweet buns flavoured with saffron and orange blossom; candied orange and lemon slices; creamy confections like *Raspberry charlotte* (see recipe) or nutty ones like fig and almond roll.

Produce of sea and soil

Provençal cooking uses many varieties of fish – from the little anchovy upwards, succulent young fruit and vegetables, olive oil, herbs and wines. At its simplest there are the unique provençal ways with grilled fish: brushing the fish with a branch of thyme soaked in olive oil, or flaming it with dried fennel. On a more sophisticated level are the preserved fruit of Apt, Avignon and Nyons, the almond sweets (*calissons*) of Aix and the nougats of Montelimar.

Other specialities include the pungent garlic mayonnaise, *aïoli*, served as an accompaniment to classic soups like *bouillabaisse* and *bourride* and also to meat and raw vegetables (*crudités*). Like *bouillabaisse*, stuffed tripe is a dish associated with Marseilles, France's biggest and most ancient port. Typical of provençal charcuterie are the delicious smoked sausages of Arles which are made of a mixture of pork and donkey meat. Another regional 'celebrity', *brandade de Nîmes*, consists of poached, creamed salt cod.

Lavender fields and olive trees are a common sight in Provence – and the backbone of the area's economy

Niçoise onion flan

● **Preparation: 1 hour, plus rising**

● **Cooking: 1 hour 20 minutes**

40g/1½oz butter
salt and pepper
175g/6oz flour
½tsp sugar
1tbls easy-blend dried yeast
1 egg, well beaten
oil, for greasing
700g/1½lb onions, thinly sliced
75ml/3fl oz olive oil
1 red pepper, seeded (optional)
dried basil
1 garlic clove, finely chopped
10 canned anchovy fillets
10 black olives, stoned

● **Serves 4-6**

● **475cals/1995kjs per serving**

1 In a large bowl, cut the butter into ½tsp salt and the flour and rub together. Stir in the sugar, then the yeast. Beat the egg with 2-3tbls warm water and pour into a well in the flour. Combine the ingredients and knead for 1 minute.

2 Roll the dough into a ball, put into a well-oiled bowl and make a cut in the top of the dough. Cover the bowl and leave it in a warm place to rise for 2 hours or until doubled in size.

3 Meanwhile, make the filling. Put the onions into the oil in a large frying pan over very low heat. Turn them with a wooden spoon until they are very oily, cover and let the onions become very soft (30-40 minutes), turning them occasionally to keep them from browning.

4 If using the pepper, heat the grill to medium. Grill the pepper, turning, until its skin blackens and bubbles. Peel off the skin and chop the flesh.

5 When the dough has risen, turn it onto a floured board, knead for 1 minute and roll it into a ball. Oil a 20cm/8in round sandwich tin, place the ball of dough in the centre and press it outwards and upwards with your fingers to fit.

6 Season the onions with salt and pepper and dried basil to taste, stir in the chopped garlic and press the mixture into the flan. Arrange the anchovy fillets like the spokes of a wheel on top, with the olives between them. Sprinkle the pepper on top. Heat the oven to 200C/400F/gas 6.

7 Let the flan rise again for 15 minutes, then set it on a baking sheet in the centre of the oven and bake for 20 minutes. Turn the heat down to 180C/350F/gas 4 and bake for a further 15-20 minutes. Serve warm or cold.

Bouillabaisse

- **Preparation: soaking the shellfish, then 1½ hours**

- **Cooking: 1 hour**

450g/1lb each of 6 of the following: mussels, langoustines, monkfish, John Dory, weaver, sea bass, red mullet, rascasse, prawns, cod, whiting
⅛tsp ground saffron
For the stock:
200ml/7fl oz olive oil
1 large onion, sliced
100g/4oz shallots, sliced

FISHERMEN'S JOY

Bouillabaisse, the famous French fish soup, is traditionally associated with Marseilles. It was originally cooked on the beach by fishermen, over a wood fire so that the flavour of smoke was absorbed. The sailors threw 'one for the pot' into the cauldron as they fished – usually fish that was least suitable to be sold at the market.

Opinions are divided on what ingredients an 'authentic' bouillabaisse should have but olive oil, saffron and orange or lemon zest seem to be essential flavourings.

450g/1lb large, ripe tomatoes, chopped
3 garlic cloves, finely chopped
1kg/2¼lb fish trimmings from the above, plus extra if necessary
3 large sprigs of tarragon
3 large sprigs of thyme
small bunch of parsley
2 bay leaves
strip of lemon zest
salt and pepper
For the rouille sauce:
2 small red peppers, seeded and finely chopped
2 garlic cloves, finely chopped
50g/2oz fresh white breadcrumbs
75ml/3fl oz olive oil
½tsp chilli powder

- **Serves 6-8**

- **770cals/3235kjs per serving**

1 Scrub the mussels, if using, discarding any that are still open. Soak for 2 hours with the langoustines, if using. Trim the fish and cut into 4cm/1½in chunks. Reserve the trimmings. Cover and chill the prepared fish until needed.

2 Make the stock. In a large saucepan, heat the oil and sauté the onion and shallots for 8 minutes, stirring frequently.

Stir in the tomatoes and garlic and cook for 5 minutes, stirring. Add the fish trimmings, herbs, lemon zest and 2L/3½pt water. Bring to the boil and simmer, uncovered, for 30 minutes. Strain, then return to the rinsed-out pan. Discard the trimmings and herbs. Season to taste with salt and pepper.

3 Meanwhile, prepare the rouille sauce. Combine the red peppers, garlic, breadcrumbs, 6tbls of the fish stock and the olive oil in a blender or food processor. Season with chilli powder and salt to taste. Transfer to a serving bowl.

4 Bring the strained stock back to the boil. Add the fish and saffron and simmer for 10-15 minutes, or until the white fish is just tender and the shellfish is cooked through. Pour into warmed serving bowls and serve immediately. Top with rouille sauce or hand it around separately.

Serving ideas

Put three slices of toasted French bread in the bottom of each warmed soup bowl. Ladle the soup onto the bread. You can also top each serving with aïoli or serve some separately (aïoli is made by adding one or two cloves of crushed garlic to mayonnaise).

Squid niçoise with olives

- **Preparation: 50 minutes**

- **Cooking: 1 hour 10 minutes**

450g/1lb baby squid
seasoned flour
oil, for deep frying
parsley sprigs, to garnish
For the niçoise sauce:
2tbls olive oil
1 Spanish onion, chopped
2tbls flour
150ml/¼pt red wine
400g/14oz canned tomatoes
3 garlic cloves, finely chopped
1 bouquet garni
1 tsp sugar
salt and pepper
1tbls capers
15 black olives, stoned

- **Serves 4**

- **355cals/1490kjs per serving**

1 To make the niçoise sauce, heat the olive oil in a saucepan and cook the chopped onion for 1-2 minutes or until ▶

transparent. Sprinkle with flour and cook for 1-2 minutes more, stirring constantly. Gradually stir in the wine and bring to the boil, stirring constantly. Add the tomatoes with their juices, the garlic, bouquet garni and sugar. Season to taste with salt and pepper. Cover and leave to simmer for 1 hour or until the sauce has thickened, stirring occasionally.

2 Meanwhile, prepare the squid. Remove the heads from the squid. Empty the cavity and remove the blade from each one. Peel the thin black membrane from the outside of the flesh and wash the squid under cold running water, checking that the insides are thoroughly cleaned.

3 With a sharp knife, cut the flesh into 5mm/¼in wide rings. Dry the rings carefully with absorbent paper. 🕐

4 Heat the oil in a deep-fat fryer to 190C/375F or until a 1cm/½in square of day-old white bread turns golden brown in 50 seconds. Just before frying the squid rings, add the capers and olives to the sauce and cook for a few minutes to heat through.

5 Coat the squid rings with seasoned flour, put them in the frying basket and fry in the hot oil for 1 minute or until golden (fry in two batches if your frying basket is not large enough). Drain on absorbent paper.

6 Spoon the niçoise sauce over the base of a serving plate and place the squid on top. Garnish with parsley.

Boeuf en daube

- **Preparation: 1¼ hours, plus cooling**

- **Cooking: 6½ hours**

2 large onions, coarsely chopped
4tbls olive oil
1tbls tomato purée
1 onion, wrapped with a strip of orange zest held in place by 5 cloves
2 large carrots, split lengthways
4 garlic cloves
200g/7oz streaky unsmoked bacon, cut across into strips
1 calf's foot or pig's trotter, split lengthways, or the knuckle of a pork or gammon joint
1.1kg/2½lb braising steak, cut into chunks about 75g/3oz each
a bouquet garni made from a sprig

of thyme and parsley and a bayleaf tied together with string
600ml/1pt or more red wine
salt and pepper

- **Serves 6**

- **660cals/2770kjs per serving**

1 Heat the oven to 150C/300F/gas 2. Cook the onions in the oil over moderate heat in a heavy flameproof casserole with a close-fitting lid until soft. Stir in the tomato purée. Continue cooking for 5 minutes, then remove from the heat and allow to cool for 5-10 minutes.

2 Place the orange-wrapped onion in the centre of the casserole. Lay the carrots over the bottom. Distribute the garlic cloves evenly, and sprinkle the bacon strips over them. Arrange the two halves of calf's foot, pig's trotter or knuckle on either side of the onion and place the meat chunks all around them, tucking the bouquet garni into the centre. Pour in enough red wine to cover and sprinkle with plenty of pepper.

3 Cover the dish closely with foil and then with a lid and bring it to the boil over a moderate heat. Reduce the heat and simmer for about 10 minutes.

4 Put the casserole in the oven for 6 hours, checking towards the end that the amount of wine is adequate. 🕐 Just before serving, add salt, and more pepper to taste.

OLIVE SECRETS

Both green and black olives come from the same tree and are one and the same fruit — green olives are simply unripe ones. Green olives are usually gathered from the end of August, whereas the finest black olives are harvested in the coldest winter months.

Ratatouille

- **Preparation: 40 minutes, plus degorging**
- **Cooking: 1 hour 20 minutes**

3 large aubergines, halved
   lengthways
3 large courgettes, halved
   lengthways
salt and pepper
150ml/¼pt olive oil
3 large onions, sliced into rings
4tbls tomato purée
4 garlic cloves, chopped
3 large green or red peppers, seeded
   and cut into thin strips
5 large tomatoes, skinned and
   chopped, or 400g/14oz canned
   tomatoes, well drained
pinch of ground coriander
small pinch of ground cinnamon
pinch of dried basil

- **Serves 6-8**
- **265cals/1115kjs per serving**

1 Cut the aubergines and courgettes across into slices about 2cm/¾in thick. Place them in layers in a colander, sprinkling each layer with salt. Top them with a weighted plate and leave to degorge for 1 hour.

2 Heat the oil in a broad, heavy pan over low heat and cook the onions in it for about 15 minutes or until trans-

AÏOLI CELEBRATION

The pungent garlic mayonnaise, *aïoli,* is so much part of provençal cooking that the traditional Friday (a fast day) meal was named after it. The creamy-yellow sauce accompanies poached salt cod, boiled potatoes and carrots at many family luncheon tables every Friday. The *grand aïoli* is a rather special celebration with additional elements like artichokes, snails, asparagus and chickpeas — only exceeded by the *aïoli monstre,* a celebration on a large scale organized each summer in villages throughout Provence.

parent. Stir in the tomato purée and cook for 3-4 minutes, stirring occasionally.

3 Rinse, then dry the aubergines and courgettes with absorbent paper and stir them into the pan. Add the garlic and the peppers, shake the pan, cover and let it simmer for about 20 minutes.

4 Add the tomatoes, spices, basil and pepper to taste. Stir once or twice and cook for a further 40-45 minutes. Remove the lid for the final few minutes to let the sauce reduce if the dish seems too liquid. Check the seasoning again just before serving.

Variations

The proportions of ingredients in Ratatouille can be varied according to taste. It is not uncommon to omit the courgettes or reduce the amount of garlic. Ratatouille can also be served cold as an hors d'oeuvre.

Salade niçoise

- **Preparation: boiling the eggs, then 40 minutes**
- **Cooking: 20 minutes**

225g/8oz new potatoes
salt and pepper
225g/8oz French beans
90g/3½oz canned tuna in oil,
   drained and lightly flaked
1 large, crisp lettuce heart
50g/2oz canned anchovy fillets,
   drained
2 hard-boiled eggs
4 firm, ripe tomatoes, skinned and
   seeded
12 black or green olives, stoned
2tbls freshly chopped mixed herbs

▶

THE WINES OF PROVENCE

The scents of the local herbs are – mysteriously – reflected in the bouquets of the wines which should not be dismissed as universally rough. Châteauneuf-du-Pape from the Rhône river is still the best known. Other provençal wines worth investigating are the Haut Comtats from Nyons, the Côteaux d'Aix, and the Tavel rosés.

◀ For the dressing:
4tbls olive oil
1tbls wine vinegar
pinch of mustard powder
1 garlic clove, crushed

- **Serves 4**

- **285cals/1195kjs per serving**

1 Cook the potatoes in lightly salted boiling water until tender, drain and leave until cool enough to handle. At the same time cook the beans in plenty of lightly salted boiling water until just tender; drain and refresh in cold water to stop the cooking process, then pat dry.

2 Slice the beans into 2.5cm/1in lengths and place in a bowl. Cut the potatoes in their skins into chunks. Add the potatoes and tuna to the beans and stir lightly to mix.

3 To make the dressing, put the oil, vinegar, mustard powder and garlic in a jar with a screw-top lid. Add salt and pepper to taste. Replace the lid tightly and shake vigorously to emulsify the ingredients. Pour a generous third of the dressing over the bean and potato mixture and stir gently but thoroughly. Set the bowl aside until the beans and potatoes are quite cold.

4 Tear the larger lettuce leaves into two or three pieces and place the lettuce in a large, fairly shallow salad bowl. Pour over half the remaining dressing and toss the leaves until evenly coated.

5 Spread the tossed lettuce over the base and sides of the bowl, then pile the bean and potato mixture in the centre. Split the anchovy fillets lengthways in half and arrange lattice fashion over the beans and potato mixture. Slice the eggs and tomatoes into wedges and arrange around the bean and potato mixture. Scatter over the olives. Sprinkle the remaining dressing over the eggs and tomatoes to moisten them. Scatter over the herbs and serve.

Raspberry charlotte

- **Preparation: 40 minutes**

- **Cooking: 5 minutes, plus chilling**

750g/1lb 10oz raspberries
100g/4oz icing sugar, plus extra for dusting
1tbls powdered gelatine
425ml/³/₄pt double cream
36 sponge fingers or boudoir biscuits
extra raspberries, to garnish (optional)

- **Serves 6-8**

- **480cals/2015kjs per serving**

1 Put 250g/9oz of the raspberries, 2tbls icing sugar and 3tbls water in a small saucepan and cook over low heat for 5 minutes. Press the mixture through a nylon sieve, sprinkle the gelatine on top of the still-hot liquid, whisk, then allow it to stand until dissolved. (If it doesn't dissolve place over a pan of simmering water until it does.)

2 Whip the cream until thick. Reserve 2tbls in a piping bag fitted with a rosette nozzle for decorating, if wished. Gradually and gently fold in first the remaining sugar, then the gelatine mixture, then the remaining raspberries.

3 Line the bottom of a 1.1L/2pt charlotte mould with a round of greaseproof paper, then use the sponge fingers, sugar sides out, to line the bottom and sides. Spoon the fruit and cream mixture into the centre. Chill for at least 6 hours.

4 Unmould the charlotte and decorate the centre with a rosette of cream and extra raspberries, if wished. Dust with a little icing sugar before serving.

Fruits of the Earth

Truffles, chestnuts and delicious full-bodied wines are just

some of the delicacies that make this part of France

a gourmet paradise

Roast pork (pages 64-5)

O VER IN THE south-west of France lie the vineyards of Bordeaux and the fertile lands of Périgord where some of the finest food in Europe is prepared.

Wine capital of the world

The Romans were the first to plant vines in the Bordeaux area and the English made claret one of the world's favourite wines – indeed, it was their development of the local wine trade that brought great prosperity to the province. It was natural that local cooking should seek to reflect the excellence of such favourites as Médoc, St Emilion and Graves. The famous *sauce bordelaise* which sits so well on grilled steak (page 65) is a case in point: ideally, the best Bordeaux should be used in its preparation.

Truffle hounds

The truffle is a rare and costly fungus which lends an unusual and delicious aroma to Périgord specialities such as *pâté de foie gras* – which was invented in this province – and *sauce Périgueux*. They used to be unearthed by specially-trained muzzled pigs but nowadays dogs do the job just as efficiently. Only the smallest amount is needed to add the distinctive flavour that makes the truffle an unmistakable ingredient. The fragrant wild mushrooms known as ceps grow abundantly here and can be used instead of ordinary mushrooms for an unusual stuffing for baked trout (page 65).

Nuts abounding

Chestnuts grow all over this area and are used to flavour both meat and vegetables as well as to make delectable *crème de marrons* and *marrons glacés*. Walnuts are cultivated here too and their light, delicious oil is widely used in salad dressings instead of olive oil. When steeped in *eau de vie*, they produce a powerful liqueur.

Picking the grapes in the vineyards near Bordeaux

Gironde onion soup

This version is a bit richer and more refined in flavour than the traditional French onion soup.

● **Preparation: 25 minutes**

● **Cooking: 25 minutes**

6 large or 12 medium-sized onions
75g/3oz lard
good pinch of dried thyme
1tsp salt
¼tsp freshly ground black pepper
1.5l/2½pt light chicken or veal
   stock
3 medium-sized egg yolks
½tsp white wine vinegar
6 slices French bread, crisped in the
   oven and rubbed with ½ garlic
   clove

● **Serves 6**

● **245cals/1030kjs per serving**

1 Peel the onions and slice them into thin rounds. Soften them in the melted lard in a covered saucepan over very low heat for 15-20 minutes, stirring from time to time to avoid browning.

2 When they are soft and translucent but not brown, add the thyme, salt and pepper. Shake the saucepan, add the stock and bring it to the boil. Lower the heat and simmer, covered, for 15-20 minutes.

3 Beat the egg yolks until smooth and pale, add the vinegar and beat some more. Gradually whisk in one ladleful of the hot soup.

4 Remove the saucepan of soup from the heat and stir in the egg mixture. Keep the soup hot without boiling again, check the seasoning and serve garnished with the crisp French bread slices.

VINES AND WINES

Wherever you travel in the Bordeaux area, you will find the great claret-producing vineyards. It is only when you are far away from the city that you will find market gardens and meadows for fruit and vegetable farming. The Périgord area too has its wines – Cahors, Bergerac and sweet white Montbazillac are just a few examples.

Roast pork with chestnuts

● **Preparation: 35 minutes**

● **Cooking: 2 hours**

1kg/2¼lb boned loin of pork,
   rolled and tied
2 garlic cloves, thinly sliced
15g/½oz lard
1 large onion, chopped
225g/8oz dried chestnuts
salt and freshly ground black
   pepper

- *Serves 4* 🍴 £££

- *640cals/2690kjs per serving*

1 Heat the oven to 190C/375F/gas 5. Pierce the pork at intervals with the point of a knife and insert the garlic slices.

2 Melt the lard over brisk heat in a heavy shallow ovenproof pan and brown the pork on all sides. Tip the chopped onion around the pork, moisten with a little water, and roast in the oven for 1½ hours.

3 Meanwhile, put the chestnuts in a saucepan with cold water, bring to the boil and simmer for 25-30 minutes, then drain. About half an hour before the roast is done, add the chestnuts to the pan, together with a little of their cooking liquid. Season with salt and pepper and finish cooking.

4 Transfer the pork to a warmed serving dish. Using a slotted spoon, drain the chestnuts and onions of any excess fat and arrange them round the roast.

Mushroom-stuffed trout

- *Preparation: 15 minutes*

- *Cooking: 40 minutes*

125g/4½oz dried mushrooms
4 fresh trout, 250g/9oz each, cleaned but with the heads left on
75g/3oz smoked back bacon, finely chopped
12 sprigs fresh parsley, chopped
5tbls chopped chives
4-5 fresh sorrel leaves, chopped
1tbls cooking oil
3tbls dried breadcrumbs
salt and pepper to taste
1 egg, beaten
a little oil
a dash of white wine vinegar
parsley to garnish

- *Serves 4* 🍴 £££

- *425cals/1785kjs per serving*

1 Pour tepid water over the dried mushrooms (preferably ceps) and leave them to soak for 20 minutes. Heat the oven to 190C/375F/gas 5.

2 While the mushrooms are soaking, chop the bacon, parsley, chives and sorrel leaves. Rinse the mushrooms in fresh water after soaking, pat them dry with kitchen paper, and chop them.

3 Heat the oil gently in a small saucepan, add in the bacon pieces and the mushrooms, and fry them over low heat, stirring constantly for 5-8 minutes. Remove the pan from the heat and stir in the chopped herbs and the breadcrumbs. Season well with salt and pepper and bind the mixture with the beaten egg.

4 Spoon the stuffing into the cavities of the fish and sew the edges together to hold it in place. Brush the outsides of the trout with oil and arrange them on an oiled baking tin large enough to take the stuffed trout in one layer.

5 Put the trout in the oven and cook for 20-25 minutes until the skins are bubbling. Transfer them to a warmed serving dish. Add a dash of vinegar and a little salt to the pan juices, stirring well. Pour this liquid over the fish. Remove the threads, garnish with parsley sprigs and chopped parsley and serve at once.

Grilled steak with red wine sauce

Most good butchers have marrow bones and will saw one into manageable pieces from which marrow – essential for the flavour of the sauce – is easily extracted.

- *Preparation: 20 minutes, plus 5 hours soaking marrow*

- *Cooking: 24 minutes*

2 entrecôte steaks about 2.5cm/1in thick
2tsp beef marrow
few drops of olive oil
salt and freshly ground black pepper
50g/2oz shallots, finely minced
100ml/3½fl oz red wine
1tsp chopped parsley
100g/4oz butter, at room temperature
few drops of lemon juice (optional)

- *Serves 2* 🍴 ££

- *735cals/3085kjs per serving*

1 Soak the marrow-bone segments for 5 hours in cold water. Drain, cover with fresh cold water, put over medium heat and poach for about 15 minutes, never ▶

allowing the water to boil. Drain again, and when cool enough, scoop out the marrow with a round-ended knife or small teaspoon.

2 Lightly brush the steaks with oil; salt and pepper them on both sides and reserve while preparing the sauce.

3 Heat the grill to medium-high. Combine the shallots and wine in a tiny pan and cook over low heat until the shallots are completely soft and the wine has almost evaporated. Remove the pan from the heat, stir in the marrow and the parsley and reserve.

4 Grill the steaks 3 minutes per side for rare, 4 minutes per side for medium. Meanwhile, cut the butter into little pieces and beat them one by one into the still-warm (but not hot) shallot and wine mixture so that it blends into a creamy sauce without completely melting.

5 Season the sauce with a few drops of lemon juice, if desired, and salt. Strain the sauce over the steaks; serve at once.

MARIE BRIZARD

A favourite French liqueur is anisette and a popular brand is based on a recipe brought from the West Indies in the 18th century by Marie Brizard, a nurse from the city of Bordeaux. Try it as an unusual addition to sweet pancakes (anisette crêpes, opposite) – or let a small glass of it help your digestion after a hearty meal.

Anisette crêpes

The addition of Marie Brizard anisette, the local aniseed-flavoured liqueur, gives this dessert a unique flavour.

● *Preparation: 20 minutes, plus 2 hours standing*

● *Cooking: 40 minutes*

300g/11oz flour
100g/4oz caster sugar
pinch of salt
1 large egg
350ml/13fl oz milk
1tbls oil
3¹/₂tbls anisette
oil or lard for greasing

● *Serves 4*

● *515cals/2165kjs per serving*

1 Combine the flour, sugar and salt in a bowl and make a well in the centre.

2 Beat the egg and milk together and pour them into the well. Add the oil and then the anisette, gradually whisking them all together with a fork. Continue whisking until all the flour is incorporated into the thin-creamy liquid. Cover with a clean cloth and leave the batter to rest for 2 or more hours.

3 Transfer the batter to a jug for easier pouring. Heat a 12-15cm/5-6in diameter frying-pan, then carefully rub it with a thick wad of absorbent paper smeared with oil.

4 Pour in 1-2tbls of batter (enough to thinly coat the bottom of the pan) and place the pan over medium-high heat. When little air bubbles start to form on the surface, flip it over with a flexible knife blade. Do not overcook. Keep warm until the rest are cooked. Serve at once.

Cook's tips

To keep crepês warm, store over hot water on a plate. In a fridge they keep for days in an airtight plastic bag. To freeze, stack them with piece of lightly oiled paper between.

Sea-swept Brittany

A country set apart from its neighbours by its ancient customs, Brittany produces delectable seafood and the finest fruit and vegetables

Baked oysters (page 70)

BRITTANY WAS ONCE Armorica, a country separate from France with its own rulers, language and customs. Even today, the Bretons are fiercely proud of their traditions and their common Celtic ancestry with Wales and Cornwall.

Fishing for history

Bathed on three sides by the Atlantic, it has a long history of seafaring and fishing and its finest dishes are based on the fresh and succulent seafood brought daily into harbour. One of the great dishes of the world is the local lobster cooked with shallots, white wine and brandy or calvados, known locally as *homard à l'armoricaine* – though many cooks prefer to call it *homard à l'américaine.*

Seafood paradise

Gourmets fall silent before the splendour of the Breton *plateau de fruits de mer*, a tantalising selection of oysters, sea urchins, *praires* and *palourdes* (tiny sweet clams). There

will be cooked shellfish too – mussels, winkles, prawns, langoustines and shrimps. You could drink a crisp Muscadet or Gros-Plant with this. Local fish includes sea bass, monkfish, conger eel and mackerel and these are included in the Breton fish soup, *cotriade*. This is often served as two separate courses, first the soup and then the pieces of fish.

Inland lie the dairy regions and in France the salty Breton butter is sometimes preferred to its rivals from Normandy and the Charentes. Breton vegetables are famous for their variety and quality, especially the artichokes and cauliflowers. Local markets are piled high with produce, especially the onions and garlic brought over to Britain by the *marchands d'oignons* with their berets and bicycles.

Perfect pancakes

In every market there will be a *crêperie* where you can eat delicious thin lacy *crêpes dentelles* with butter and sugar or jam, or the thicker buckwheat *galette* served stuffed with savoury items such as cheese, ham and onions. Drink local cider with these and remember that it can pack quite a punch!

Crab and lobster fishermen at Roscoff, Brittany

Breton fish with vegetables

● **Preparation: 50 minutes**

● **Cooking: 30 minutes**

100g/4oz smoked bacon, rind removed, diced
100g/4oz butter
3 large shallots
100g/4oz diced carrots
100g/4oz diced baby turnip
225g/8oz celeriac, cut in large cubes
400g/14oz new potatoes, grated
225g/8oz green peas
salt and freshly ground black pepper
1kg/2¼lb whole firm white fish such as a fresh haddock or monk fish, cut into 4 portions with head and tail reserved
250ml/9fl oz dry white wine
beurre manié made with 15g/½oz butter and 15g/½oz flour
150ml/5fl oz thick cream, lightly whipped
For the fish stock:
1 carrot, sliced
1 turnip, sliced
1 onion, sliced
1 garlic clove, crushed
1 clove
200ml/7fl oz dry white wine
2 tomatoes, skinned and seeded
1 bouquet garni
salt and freshly ground black pepper

● *Serves 4*

● *875cals/3675kjs per serving*

1 Place all the ingredients for the fish stock in a large pan and pour in 600ml/1pt water. Bring to the boil, cover and simmer for 30 minutes.

2 Heat the oven to 220C/425F/gas 7. Sauté the bacon dice in a saucepan over gentle heat until the fat runs. Add the butter and when it melts put in the shallot, carrot, turnip, celeriac and potato. Stir well, cover the pan and cook over moderate heat for about 20 minutes, stirring occasionally, or until the vegetables are tender. Add the peas and season lightly with salt and pepper.

3 Put the vegetable mixture in a shallow ovenproof dish and bed down the pieces of fish among the vegetables.

4 Pour the wine into the pan in which the vegetables were cooked and stir over moderate heat for 1 minute. Strain in the fish stock and bring to the boil. Reduce the heat and boil until reduced by half.

5 Pour half the reduced stock over the fish and cook uncovered in the oven for 25 minutes, basting the fish with juices in the dish once during this time.

6 Meanwhile, knead together the butter and flour to make the beurre manié.

7 Heat the remaining reduced stock until boiling. Add the beurre manié in small pieces, stirring continuously until the sauce is smooth and thick. Simmer for 2 minutes then blend in the cream and reheat but do not allow the sauce to boil. Pour over the fish just before serving.

Brittany scallop gratin

Coquilles Saint-Jacques à la bretonne is an excellent hot *hors d'oeuvre* which can be prepared in advance and slipped under a hot grill to reheat and brown the top just before serving

● *Preparation: 20 minutes*

● *Cooking: 20 minutes*

8 large scallops, fresh or frozen
150ml/5fl oz dry white wine
salt and freshly ground black
   pepper
75g/3oz stale white breadcrumbs
6tbls milk
1 Spanish onion, finely chopped
1 shallot, finely chopped
75g/3oz butter
1tbls cognac
1 garlic clove, crushed
2tbls finely chopped parsley
1tbls flour
4tbls fine, fresh white breadcrumbs
2tbls finely grated cheese,
   optional

● *Serves 4*

● *340cals/1430kjs per serving*

1 Remove the scallops from the shells and trim off the valves. Rinse the scallops to remove all traces of sand. If using frozen scallops, allow them to thaw.

2 Cut each scallop into large dice, leaving the coral in one piece.

3 Place the dried scallops and corals in a small pan; add the wine and salt and freshly ground black pepper to taste. Bring the wine to a simmer, poach very gently for about 5 minutes. Drain, reserving the poaching liquor and keep hot.

4 Put the stale breadcrumbs in a bowl and sprinkle with the milk.

5 In a small saucepan, sauté the finely chopped onion and shallot in 50g/2oz of the butter until soft and golden. Add the poached scallops and continue to sauté for 2-3 minutes, stirring gently with a wooden spoon. Stir in the cognac.

6 Add the soaked breadcrumbs, together with the crushed garlic and finely chopped parsley, and simmer for 2-3 minutes longer, stirring constantly.

7 Sprinkle the flour into the pan and moisten the mixture with the reserved poaching liquor. Season to taste with salt and freshly ground black pepper and allow to simmer for a final 5-6 minutes, stirring. The sauce should have the consistency of a béchamel. Heat the grill to high.

8 Scrub 4 curved scallop shells clean and sterilise in a solution of Milton, rinse in cold water. Or use individual flameproof ramekins instead. Divide the scallops and the sauce equally between the shells. Sprinkle with breadcrumbs and some of the parsley, retaining some for garnishing, dot with the butter.

9 Just before serving, place the shells under the grill until the tops are golden brown. If using, mix cheese with breadcrumbs before grilling.

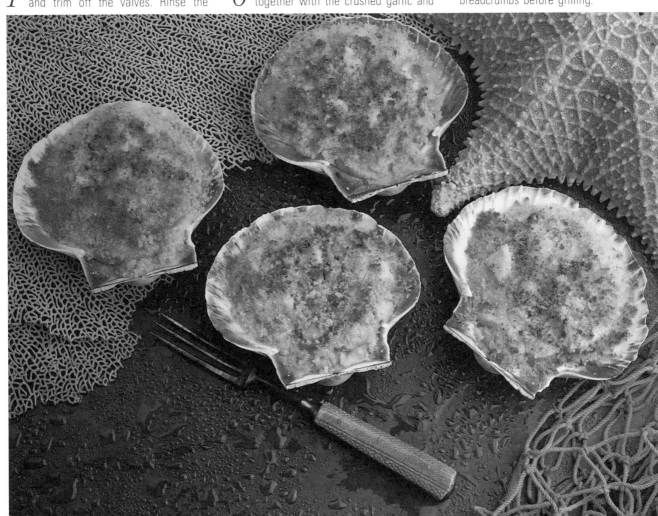

Baked oysters

The Bretons serve oysters raw on a bed of seaweed laid over crushed ice, accompanied by bread, butter, slices of lemon, or vinegar seasoned with finely chopped shallots

- *Preparation: 30 minutes*

- *Cooking: 20 minutes*

28 oysters
about 250ml/9fl oz Muscadet or
*   other dry white wine*
beurre manié made with 15g/½oz
*   butter and 15g/½oz flour*
2tbls thick cream
1 egg yolk
salt and cayenne pepper
2tbls finely chopped parsley

- *Serves 4 as starter*

- *160cals/670kjs per serving*

1 Heat the oven to 220C/425F/gas 7. Open the oysters and remove the flesh, reserving the shells. Carefully strain the liquid through a fine sieve into a measuring jug. Make the liquid up to 300ml/½pt with wine.

2 Put the oysters covered by the mixture in a saucepan over low heat and cook just below a simmer for 1-2 minutes, according to size. Carefully lift out the oysters with a slotted spoon on to a flat dish and cover until needed.

3 Reduce the cooking liquid by about one third and remove from heat. Work the butter and flour together to make beurre manié and add it to the pan in small pieces, stirring briskly and continuously over moderate heat until the sauce is thick and smooth. Gradually stir in the cream. Remove from the heat, allow to cool slightly and beat in the egg yolk. Season with salt and cayenne to taste.

Saddle of lamb with flageolets
(page 71)

ALL SORTS OF CIDER

Breton apple orchards produce the fruit for a wide variety of ciders, some mild and delicate enough for family drinking and some strong enough to make the head spin – so watch out when you decide to indulge in the local produce! Perry, made from local pears, is often called the Breton champagne. The flavour is prized for its subtlety.

4 Choose the 24 best-shaped oyster shells, wash, dry and butter them. Place one oyster in each and spoon over a little of the sauce. Arrange them on a baking sheet and bake for 10 minutes. Sprinkle all over the tops of the shells with the chopped parsley.

LAMB WITH A DIFFERENCE

Breton sheep graze on the salt meadows by the sea (the *prés salés*) and are considered to have a specially delicate flavour. Large joints such as gigot or leg and the saddle are often served with dried beans. The saddle can be marinated in local cider and served with beans (page 71).

Saddle of lamb with flageolets

Lamb in Brittany has a special flavour due to the slightly salty grass of the region. Large joints such as a *gigot* or leg, and the saddle, are usually served with a dried bean accompaniment. Canned flageolets can be substituted for dried ones. Just drain and heat before using

- ● *Preparation: 35 minutes, plus 12 hours marinading*

- ● *Cooking: 2½ hours*

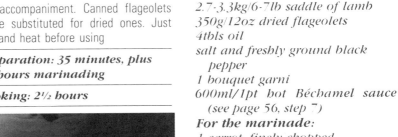

2.7-3.3kg/6-7lb saddle of lamb
350g/12oz dried flageolets
4tbls oil
salt and freshly ground black pepper
1 bouquet garni
600ml/1pt hot Béchamel sauce (see page 56, step 7)

For the marinade:
1 carrot, finely chopped
1 large onion, finely chopped
2 sprigs of parsley
1 sprig of thyme
1 bay leaf
4 peppercorns
2 cloves
4tbls cider vinegar
300ml/½pt dry cider
2tbls brandy

- ● Serves 8

- ● 870cals/3655kjs per serving

1 To make the marinade, put the prepared vegetables in a pan with the parsley, thyme, bay leaf, peppercorns, cloves and vinegar. Add 300ml/½pt water and bring to the boil. Cover and simmer for 20 minutes then leave to cool. Stir in the cider and brandy.

2 Place the saddle of lamb in a large plastic bag and pour over the marinade. Leave to stand in a dish for 12 hours, refrigerated, turning the joint several times. Soak the beans in cold water to cover for the same period.

3 Heat the oven to 180C/350F/gas 4. Melt the lard in a roasting tin. Remove the joint from the marinade and place it in the tin, skin side upwards. Reserve the marinade. Roast on the lower shelf of the oven for 2½ hours, basting occasionally. If the meat is preferred slightly pink, allow a shorter roasting time.

4 As soon as the joint is in the oven, drain the beans and put them in an ovenproof casserole with a little seasoning and the bouquet garni. Pour over boiling water to cover the beans by 20mm/¾in. Cook, covered, in the hottest part of the oven until the meat is done. If necessary, add a little more boiling water during cooking from time to time to prevent the beans from becoming dry.

5 Discard the bouquet garni and drain any remaining liquid from the beans. Transfer to a serving dish. Pour the hot béchamel sauce over them and stir gently.

6 Put the saddle on a carving board. With the chump furthest away from you, and with a very sharp knife so that you do not spoil the appearance of the crispy fat, cut down each side parallel, and

as close as possible, to the backbone. This divides the saddle into 4 parts.

7 Working down against the bones release these 4 parts leaving the clean bone. Turn the bones over and carve off any bits of meat.

8 Place the bones on a heated serving dish. Cut the chump ends into slices obliquely parallel to the original cuts. Cut the remaining part vertically into slices. Reassemble carefully onto the bones using the pieces carved from the underneath to hold the meat in place, and serve.

Calvados butter crêpes

- **Preparation: 20 minutes, plus 2 hours standing time**

- **Cooking: 20 minutes**

75g/3oz flour
½tsp salt
2 large eggs, beaten
2tbls melted butter or oil
150ml/5fl oz milk
unsalted butter or oil for greasing
2tbls icing sugar

For the Calvados butter:
50g/2oz unsalted butter, softened
4tbls caster sugar
3tbls Calvados

- **Serves 4**

- **385cals/1615kjs per serving**

1 Sift the flour and salt into a bowl. Stir in the beaten eggs with the melted butter or oil. Gradually add the milk, stirring until smooth. Strain the batter through a sieve and stand for 2 hours.

2 Prepare the Calvados butter; beat the softened butter in a bowl with a wooden spoon. Add the caster sugar and beat until the mixture becomes very light and fluffy. When it is almost white, add the Calvados a little at a time, beating vigorously with a wire whisk. Scrape the butter into a mound and chill until firm.

3 Heat a 17.5cm/7in frying pan, then grease it. Add 3tbls batter and tilt the pan to thinly coat the surface. Cook the crêpe for 1 minute, until bubbles start to form underneath.

4 Slip a palette knife underneath the crêpe and turn it over. Cook for 1 minute on the second side. Make 7 more crêpes in the same way.

5 Heat the grill to high. To assemble the crêpes, put an eighth of the Calvados butter at one end of each crêpe. Roll up, the crêpes then fold the ends under to make a neat parcel.

6 Pack the crêpes tightly side by side in a flameproof dish placing the seams underneath on the bottom of the dish.

7 Sift the icing sugar over the top of the crêpes and place under the grill for 2-3 minutes or until glazed and golden. Transfer to individual plates and serve immediately after grilling.

Cook's tips

The filled crêpes may be left in a cool place for several hours, making this a very convenient dinner party dessert. Heat the grill to high a few minutes before serving and heat and glaze the crêpes.

CREAMERY CHEESES

Local farmers make delicious *crèmets* which have to be eaten the day they are made: they are served with sugar. Tiny curd cheeses called *caillés* can be eaten with salt or sugar and these too need to be eaten at once.

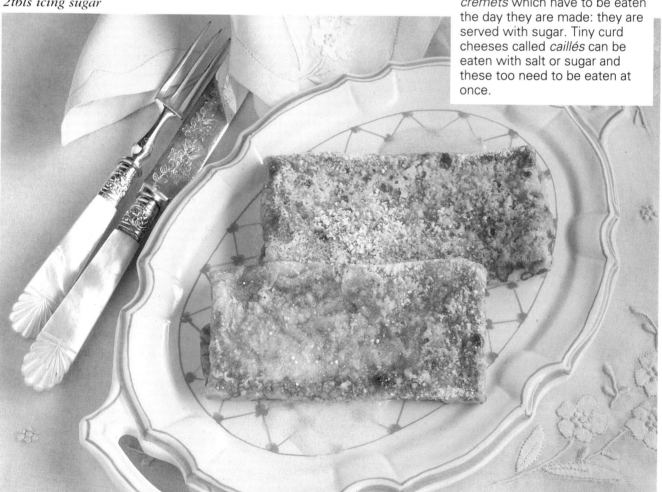

The Heart of France

Burgundy, the Auvergne, the Lyonnais and the Bourbonnais are old provinces of France whose traditional dishes have never lost their popularity

*C*LASSIC WINES FOR classic cooking – not surprisingly, the splendid wines of Burgundy have had a lasting influence on the great dishes of that region. Here, you will find red wine used in surprising ways – to cook such river fish as eels (page 77), for instance. *Sauce meurette,* a rich sauce based on a good red wine, marc de Bourgogne and vegetables, is used not only to accompany cooked chicken but also to coat poached eggs. And of course, the better the bottle of wine, the better will be your *boeuf à la bourgignonne,* that classic Burgundy stew. The great white wines have

their part to play, too: local snails are cooked with Chablis (page 74) and pork chops are simmered in white Mâcon to make a delicious mustard and wine sauce (page 78).

A wealth of onions

In cookery terms, *lyonnaise* indicates a dish that is loaded with onions, and regional specialities from around Lyons will include such fragrant and tasty items as the local version of onion soup (thickened with rice rather than cheese toast) and hard-boiled eggs served in a rich garlic, onion and cheese sauce. Many experts consider Lyons to be

Snails cooked in white wine (page 74)

the culinary capital of France and the great French chef Paul Bocuse is Lyons born and bred. You will certainly eat superbly well in this city.

Cheese and potato specials

Moving westwards into the Bourbonnais, food becomes simpler and more rustic. There is a tremendous emphasis on potatoes, especially delicious filling potato pies called *tourtons,* perfect for keeping out the cold (page 75). In the Auvergne, where the wild countryside demands food of weight and substance, the delicious local cheese, Cantal, appears in the traditional bread, cheese and onion soup, *soupe de cantal*, and in *aligot*, a warming dish of cooked whipped potatoes blended with grated cheese. This area also produces the great St Nectaire and the unusual Bleu d'Auvergne cheeses.

Working in the Vosne-Romanée vineyard in Burgundy

Snails in white wine

In the picturesque market towns, such as Beaune and Mâcon, housewives buy fresh snails and cook them at home, usually blanching the large, specially prepared snails for about 10 minutes in boiling water before removing them from the shells for cooking. Do not use garden snails; the process for cleansing them is a long one.

- **Preparation: 50 minutes for fresh – 15 minutes for canned**

- **Cooking: 40 minutes for fresh – 10 minutes for canned**

24 blanched or canned snails, drained, with the shells separate
1 garlic clove, crushed (optional)
1 small carrot, quartered (optional)
1 small onion, quartered (optional)
1 bouquet garni (optional)

MARC DE BOURGOGNE

Marc is one of those fiery spirits which can be distilled from crushed and pulped grapes after pressing – grappa is another one. They are very strong and sometimes have an odd flavour which does not appeal to everyone. Marc can be something of a gourmet pleasure, though, and it is used in Burgundy to add an indefinable something to marinades and stews. A *petit coup* after a heavy meal sits well with a cup of black coffee!

pinch of grated nutmeg (optional) (optional)
salt (optional)
3tbls Chablis or other dry white wine
For the snail butter:
150g/5oz butter
1 garlic clove, finely chopped
1tbls finely chopped shallot
1tbls finely chopped parsley
3tbls fine fresh white breadcrumbs
salt and freshly ground black pepper

- **Serves 4**

- **245 cals/1030 kjs per serving**

1 If using freshly prepared snails, put them in a pan with water to cover and add the garlic, carrot, onion, bouquet garni, nutmeg and a little salt. Bring the liquid to the boil, then cover and simmer for 30 minutes. Let the snails cool in the cooking liquid. Meanwhile, wash the shells and allow them to drain and dry naturally. If using canned snails, omit this step and start with the butter.

2 Heat the oven to 220C/425F/gas 7. To make the snail butter, beat together the butter, garlic, shallot, parsley and breadcrumbs until fairly smooth. Season the butter to taste.

3 Drain the cooked or canned snails and put them into the shells. Cover each one completely with snail butter to seal. Arrange them neatly in an ovenproof dish and sprinkle over the wine. Put them in the hot oven for 10 minutes, or until the butter topping sizzles.

It is easier to eat snails if you have special snail tongs and forks.

Baked carp with herbs

By tradition the dish containing the carp was sealed with a flour and water dough which was broken off and discarded when the fish was cooked. Nowadays foil, well crimped under the edge of the dish, is used and makes the recipe much simpler to prepare.

- **Preparation: 15 minutes**

- **Cooking: 1¼ hours**

butter for greasing
1 large carrot, sliced
1 large onion, sliced
2 shallots, chopped
1 garlic clove, crushed
2 bay leaves
1tsp dried thyme

1 tbls chopped parsley
salt and freshly ground black
 pepper
1.5kg/3¼lb carp, scaled and
 cleaned
4tbls wine vinegar
4tbls olive oil plus extra for
 greasing
For serving and garnishing:
4 large oval croûtons, about 10cm/
 4in in length, fried in butter
parsley sprigs and lemon wedges

● **Serves 4**

● **465 cals/1955 kjs per serving**

1 Heat the oven to 150C/300F/gas 2. Generously grease a shallow oven-proof dish just large enough to take the fish and vegetables comfortably and arrange the vegetables and herbs in it. Sprinkle lightly with salt and pepper. Lay the fish on top, pour over the vinegar and oil and add enough cold water to come about half-way up the fish. Cover the dish tightly with greased foil and cook the fish in the oven for 1 hour.

2 Arrange the croûtons on a warm serving platter. Lift out the fish carefully with 2 fish slices and allow to drain well. Place the fish on the croûtons and garnish generously with sprigs of parsley and lemon wedges.

3 Strain the juices from the cooking dish into a warm sauce-boat, add salt and pepper if necessary and serve the fish immediately, with the sauce handed separately in a jug.

CHESTNUT SPECIALS

Here, as in so many French country areas, chestnuts are widely used in cooking. In Lyons, you will find an unusual way of cooking oxtail – after long simmering in a special stock enriched with Beaune and marc de Bourgogne, freshly boiled chestnuts are added to the stew. Canned ones may be used, but the real thing imparts a special flavour.

Rich herbed potato pie

Sometimes this double crust pie is made without the addition of cream, and sometimes it is made with puff pastry instead of this pâte brisée.

● **Preparation: 1 hour plus 20 minutes chilling**

● **Cooking: 1½ hours**

450g/1lb flour
1½tsp salt
200g/7oz butter, plus extra for
 greasing
2 large eggs, separated
flour for sprinkling
For the filling:
1kg/2¼lb medium-sized waxy
 potatoes
salt
freshly ground black pepper
4tbls chopped parsley
2tbls chopped chervil or 2tsp dried
 chervil
175g/6oz finely chopped shallots ▶

75

150ml/5fl oz thick cream, lightly whipped

● **Serves 6** ⑪ ££

● **770 cals/3234 kjs per serving**

1 Sift the flour and salt into a bowl and rub in the butter until the mixture resembles breadcrumbs. Whisk one whole egg and one egg white lightly with 4tbls cold water, add this to the dry ingredients and mix with a round-bladed knife until the dough leaves the sides of the bowl clean. Chill for 20 minutes.

2 Meanwhile, blanch the potatoes for 8 minutes in a large pan of boiling, salted water. Drain well, and when cool enough to handle, cut them into slices.

3 Heat the oven to 200C/400F/gas 6. Roll out slightly more than half the pastry on a floured surface, then line a greased 23cm/9in flan tin or ring.

4 Arrange about a third of the potato slices in the pastry case, season lightly with salt and pepper and sprinkle over about half the herbs and shallots. Cover with another third of the potato slices, season and sprinkle with the rest of the herbs and shallots, then cover with the remaining potato.

5 Beat the remaining egg yolk with 1tbls water and brush the pastry edges. Roll out the rest of the pastry to make a lid and cover the pie. Seal the edges well. Brush with the remaining egg yolk wash and cut 4 small steam vents spaced evenly in the top of the pie. Bake for 25 minutes, then reduce oven heat to 180C/350F/gas 4 and cook for 40 minutes.

6 Remove the pie from the oven and open the steam vents sufficiently with a sharp knife to take a small funnel. Pour one quarter of the cream into each of the steam vents and return the pie to the oven for a further 10 minutes, or until the top is rich golden brown. Serve hot.

VICHYSSOISE

Since Vichy is a town in the Bourbonnais, one might assume that the famous chilled leek and potato soup originated there: not so – it was created far away in New York by a French chef who named it in honour of Vichy, the town where he was born. But the recipe *Carottes Vichy* was actually created in this pretty spa town and since carrots are thought to be excellent for liver complaints, this dish is often served here as part of the special cure diet.

Sausages with wine and cheese

Cervelas are highly seasoned meaty sausages and are available in two sizes; the larger sausages serve two people each.

● *Preparation: 10 minutes*

● *Cooking: 40 minutes*

4 small Cervelas sausages
butter for greasing
100ml/4fl oz white wine
75g/3oz Gruyère cheese, grated
flat-leaved parsley, to garnish

● *Serves 4*   ⑪ ££

● *255 cals/1070 kjs per serving*

1 Prick the sausages with a fork and poach them in simmering water just to cover for 30 minutes.

2 Meanwhile, heat the oven to 220C/425F/gas 7. Drain the sausages and cut each one in half lengthways. Arrange the pieces of sausage, cut surfaces downwards, in a greased shallow ovenproof dish. Spoon the wine over and sprinkle with the cheese. Bake in the oven for 10 minutes. Arrange flat-leaved parsley leaves between the cervelas halves to garnish, then serve the sausages hot from the ovenproof dish.

the fish, if used, and cut into similar sized lengths.

2 Heat the oil and 25g/1oz of the butter in a large shallow pan, add the pork and sauté until golden. Put in the pieces of eel and fish, if using, the onions and garlic. Continue to sauté for 2 minutes, or until the eel and fish begin to take on a pale golden colour.

3 Gently warm the marc, ignite and pour over the ingredients, shaking the pan occasionally until the flames die down. Add the wine, bouquet garni, peppercorns and salt to taste.

4 Bring the mixture to the boil, cover and simmer for about 30 minutes, or until the eel is tender. Strain off the stock and reserve for use in the sauce.

5 Melt another 25g/1oz of the butter in a clean saucepan. Add the flour and cook, stirring, for 1 minute. Gradually add the reserved stock, stirring constantly, until the sauce boils and thickens. Cook for 2 minutes.

6 Carefully lift the pieces of eel and cubes of pork into the sauce and adjust the seasoning if necessary. Reheat the mixture gently over a low heat to avoid the fish breaking up.

7 Meanwhile, melt the remaining butter, add the lemon juice, then fry the mushrooms in it until pale golden. Drain on absorbent paper.

8 Serve the fish stew in warm bowls topped with the fried mushrooms and triangular croûtons.

Cook's tips

To make the previous recipe into a substantial main dish, the sliced Cervelas sausages can be served on a bed of cooked lentils delicately flavoured with bay leaves and thyme, dotted with butter and sprinkled with salt and pepper.

Burgundy eel stew

● **Preparation: 35 minutes**

● **Cooking: 45 minutes**

1kg/2¼lb medium-sized eels or
*   half carp or pike and half eels*
1tbls olive oil
75g/3oz butter
100g/4oz salt pork, cubed
12 tiny onions
3 garlic cloves, crushed
6tbls marc de Bourgogne or brandy
1L/1¾pt Burgundy or other dry red
*   wine*
1 bouquet garni
4 black peppercorns
salt
25g/1oz flour
1tsp lemon juice
12 button mushrooms
4 triangular croûtons, to serve

● *Serves 4*

● *860 cals/3610 kjs per serving*

1 Clean and skin the eels and cut them into 5cm/2in lengths, discarding the heads. Remove the heads and tails from

Pork chops in mustard wine sauce

- **Preparation: 10 minutes**
- **Cooking: 30 minutes**

4 large pork chops, trimmed
salt
freshly ground black pepper
3tbls oil
4 shallots, finely chopped
200ml/7fl oz Mâcon blanc or other
   dry white wine
2tsp tomato purée
1tsp Dijon mustard
15g/¹/₂oz butter
sprigs of parsley, to garnish

- **Serves 4**
- **370 cals/1555 kjs per serving**

1 Season the chops on both sides with salt and pepper. Fry them in the oil over medium-low heat for about 15 minutes, turning them once during this time, until golden brown on both sides and cooked through. (Time will depend on thickness of chops.) Transfer them to a warm serving dish and keep hot.

2 Add the shallot to the fat remaining in the pan and cook gently over a low heat until soft. Add the wine and tomato purée, bring to the boil and cook over moderate heat for 2 minutes, stirring.

3 Add the mustard and butter and stir briskly until blended. Taste and adjust the seasoning if necessary, then spoon the sauce over the chops. Serve garnished with sprigs of parsley.

Pear turnovers

Under the name *chaussons aux poires*, these are sold in pastry shops. In other regions they would be filled with an apple mixture. Their local name, *piquenchagne*, is also used to describe a children's game.

- **Preparation: 40 minutes plus 3 hours macerating time**
- **Cooking: 25 minutes**

4 large ripe, firm pears
100g/4oz caster sugar
pinch of freshly ground white
   pepper
¹/₄tsp vanilla essence
3tbls dark rum
50ml/2fl oz double cream, whipped

For the pâte sucrée:
450g/1lb flour
pinch of salt
250g/9oz butter
15g/¹/₂oz caster sugar
flour for dusting
1 egg, beaten
To serve:
double cream, whipped

- **Makes 8 turnovers**
- **555 cals/2330 kjs per serving**

1 Peel, halve and core the pears. Chop the flesh finely and put it in a bowl. Sprinkle with the sugar and freshly ground white pepper, add the vanilla essence, rum and whipped double cream. Mix the ingredients, cover the bowl with stretch-wrap and macerate for 3 hours.

2 Meanwhile prepare the pastry. Sift the flour with the salt into a bowl and rub in the butter until the mixture resembles breadcrumbs. Stir in the sugar. Add 4-5tbls cold water to make a firm dough. Chill for 20 minutes.

3 Heat the oven to 200C/400F/gas 6. Divide the pastry into 8 equal portions and roll out each one on a floured surface with a floured rolling pin to a long oval shape about 18 × 9cm/7 × 3½in.

4 Spoon the pear filling onto one half of each pastry oval. Brush the pastry edges with egg, fold over to enclose the filling and press to seal the edges firmly.

5 Arrange the turnovers on a lightly floured baking sheet and prick each one about six times with a fine skewer. Brush them all over with the remaining beaten egg and bake for about 25 minutes, or until golden. Serve hot, with cream.

Hearty German Fare

As varied as its landscape, the cooking of Germany ranges from warming stews to elegant desserts. The one thing common throughout the land, however, is large helpings!

I T'S NOT ALL beer and sausages in Germany, though one could be forgiven for thinking this, as some of the best sausages in the world – and some of the best beer – come from that country. Whatever they eat, the portions in southern Germany are usually gigantic. There is *G'selchte* (smoked meats), sometimes just cut into thin slivers and accompanied by dark country bread, or cooked and served with sauerkraut and savoury dumplings, both of which are specialities of southern Germany. Sauerkraut is almost a necessity with most of the famous

German sausages, particularly with the Nüremberg sausages. The only exception – the one sausage never served with sauerkraut – is the *Weisswurst* (white sausage) of Munich which is usually accompanied by slightly sweet mustard.

The best from Baden

It has been said that some of the best cooking in Germany, if not *the* best, is found in the Baden region. Certainly some of Germany's best chefs come from there and it boasts some of Germany's best restaurants. The reasons for this concentrated excel-

Creamy rice and fruit pudding (page 83)

lence are many: its lakes and rivers yield beautiful fish, its forests are full of game and its orchards are second to none.

Above all in Baden there are wines. Cooking with wine seems only natural in a region where there is a local wine for practically every dish. Rich sauces are laced with wine, meats are simmered in wine and the heavy, heady wines of Baden complement every meal and turn even the simplest dish into a feast.

Influential neighbours

A large number of both hearty and delicate dishes come from East Germany, where the cuisine is as varied as the landscape. Here cooking is much influenced by neighbours both south and east; there is a touch of Bohemia, a faint flavouring from Poland and even a Viennese influence on the pastries. Silesia is the birthplace of one of the most elegant rice desserts ever, Creamy rice and fruit pudding (see recipe) invented by a chef for his princely master.

Other regions in eastern Germany have their local specialities, too, like *Köthener Schusterpfanne*, loosely translated as 'shoemaker's special', a rather unusual combination of pork, potatoes and small cooking pears, liberally spiced with caraway seeds (see recipe).

Berlin, too, is the home of many famous dishes: *Bockwurst*, a thick hot-dog-like sausage now sold in many countries, was the creation of a Berlin innkeeper; and Berlin's beer is justly famous, especially *Berliner Weisse*, or 'white' beer, though drinking it Berlin-style *mit Schuss* – with a shot of raspberry syrup – is an acquired taste.

Northern elegance

Writers and poets have sung the praises of Hamburg's food and its 'English' customs since the 19th century. One writer humorously put the importance of Hamburg's

famous smoked beef above that of the Protestant Reformation. Even today, afternoon tea is still served in the city's best houses in a rather stately way and plum pudding takes pride of place at many festive tables. Food in Hamburg seems to be a mixture of elegant dishes culled from all over the world with local embellishments plus a number of truly 'native' dishes such as Hamburg eel soup.

One finds delicious and potent drinks in northern Germany, such as *Pharisäer*, a distant relative of Irish coffee. 'It is all due to the bad quality of the water,' wrote a 19th century physician, 'coupled with the rich food and our damp, cold climate which makes these alcoholic drinks necessary'.

The Oktoberfest, or Munich beer festival, Bavaria

Chervil soup

- **Preparation: 10 minutes**

- **Cooking: 25 minutes**

1L/1¾pt beef stock
1tbls flour
125ml/4fl oz milk
2 egg yolks
125ml/4fl oz double cream
65g/2½oz fresh chervil, finely
   chopped
salt and pepper

- **Serves 4-6**　　　　　　

- **185cals/775kjs per serving**

1 Heat the stock to boiling point. In a large bowl, mix the flour to a smooth paste with the milk. Gradually add the hot stock to the paste until it is all absorbed, stirring constantly. Return it to the pan and simmer gently for 10-15 minutes.

2 In a large bowl, mix the egg yolks and cream together. Remove the soup from the heat and gradually pour it onto the egg yolks and cream, stirring to blend thoroughly. Add the chopped chervil and salt and pepper to taste.

3 Return the soup to the rinsed-out pan and gently reheat it, being careful not to let it boil. Serve it piping hot.

Sauerkraut with German sausages

- **Preparation: 20 minutes**

- **Cooking: 1 hour**

50g/2oz lard or dripping
1 large onion, sliced
4 frankfurters
1 Bratwurst sausage
2 red dessert apples, cored and cut
    into eighths
450g/1lb sauerkraut, drained
4 slices smoked loin of pork
4 slices smoked ham
1tbls soft brown sugar
2 bay leaves
6tbls dry white wine
2tsp juniper berries
1tsp caraway seeds
salt and pepper

- **Serves 4-5**

- **390cals/1640kjs per serving**

1 Heat the oven to 200C/400F/gas 6. Melt the lard or dripping in an ovenproof casserole dish over medium heat. When it starts to sizzle, add the onion and fry gently for 5-6 minutes or until softened and golden brown, stirring occasionally.

2 Add the sausages and apple pieces to the casserole and sauté for 6-7 minutes or until golden.

3 Add the sauerkraut, pork, ham, sugar, bay leaves, wine, juniper berries, caraway seeds and salt and pepper to taste. Stir well.

4 Cover and bake for 40-50 minutes, stirring once after 30 minutes. Serve hot, cutting the Bratwurst into four pieces.

ALL A BIT FISHY

Hamburg eel soup is really a stew rather than a soup, consisting of ham, dried fruit, vegetables, herbs and small dumplings to which the eel is added as if as an afterthought. Legend has it that the name *Aalsuppe* (eel soup) was originally *Allsuppe* – *All* meaning that it was a soup into which everything went. Somehow the name got changed to eel soup, so one was obliged to add the eel.

Potatoes with black pudding

This is a popular winter dish in Germany, particularly around Cologne. Its name, *Himmel und Erde*, meaning heaven and earth, refers to the light and dark colours of the mashed potato and black pudding respectively

- **Preparation: 25 minutes**

- **Cooking: 45 minutes**

1kg/2¼lb potatoes
salt and pepper
1kg/2¼lb cooking apples, peeled,
    cored and quartered
1tbls sugar, or to taste
1tbls oil
100g/4oz streaky bacon, cut into
    2.5cm/1in pieces
2 large onions, thinly sliced
450g/1lb black pudding, sliced
parsley, to garnish (optional)

- **Serves 4-5**

- **785cals/3295kjs per serving**

2 Bring a large saucepan of water to the boil, then push the batter through a colander into the boiling water. Cook for 3-4 minutes or until the Spätzle float, stirring occasionally. Drain on absorbent paper and serve immediately.

Cook's tips

Spätzle should be very light and delicate. Cook one and taste it first – if it is too heavy, add a little water to the batter.

You can form the batter by pressing through a colander, as above; or by using a special Spätzle cutter.

1 Boil the potatoes in lightly salted water for 20 minutes or until tender. Drain and mash. Reserve.

2 Meanwhile, cook the apples with the sugar and 3tbls water, covered, over low heat for 15-20 minutes or until soft and mushy. Add to the mashed potatoes, mix well and keep warm.

3 Heat the oil in a frying pan, then add the bacon and onions and cook over low heat for 10-12 minutes or until cooked but not brown, stirring occasionally. Remove from the frying pan with a slotted spoon and reserve.

4 Add the onions and bacon to the apple and potato mixture and mix well. Season with salt and pepper to taste. Transfer to a serving dish and keep warm.

5 Meanwhile, lightly fry the sliced black pudding for about 5 minutes, turning once. Add the black pudding to the potato mixture in a single layer on top. Serve immediately, garnished with parsley, if using.

Serving ideas

The combination of fruit, bacon and black pudding in this dish makes it particularly appropriate for a hearty breakfast. It can be served equally well as a supper dish, on its own or with fried liver for a more substantial meal.

Spätzle

These little dumplings are delicious served with meat or in soups

- *Preparation: 15 minutes*

- *Cooking: 5 minutes*

250g/9oz flour
salt and white pepper
2 large eggs, lightly beaten
250ml/9fl oz milk

- *Serves 4*　　　　　①£

- *305cals/1280kjs per serving*

1 Sift the flour, with a pinch of salt and pepper, into a bowl. Make a well in the centre, pour in the eggs, then gradually pour in the milk, stirring slowly and drawing the flour in from the sides. Beat the batter until smooth.

Pork and pear casserole

- **Preparation: 25 minutes**

- **Cooking: 2 hours**

700g/1½lb boned, rindless pork
    spare rib joint
salt and pepper
2tbls oil
700g/1½lb potatoes, sliced
700g/1½lb small cooking pears,
    peeled and halved
1tbls caraway seeds
about 500ml/18floz chicken stock
parsley, to garnish (optional)

- **Serves 4-6**

- **745cals/3130kjs per serving**

1 Rub the meat all over with salt. Heat the oil in a flameproof casserole over medium-high heat and brown the joint on all sides. Remove from the heat.

2 Arrange the sliced potatoes around the meat and position the peeled pears between the meat and potatoes. Sprinkle with the caraway seeds and season with salt and pepper to taste.

3 Pour in the stock, bring to the boil, cover, reduce the heat and simmer gently until the meat is cooked, about 1½ hours.

4 Remove the meat and continue to boil the liquid until thickened slightly. Garnish with parsley, if wished, and serve hot. Hand the sauce separately, in a warmed sauceboat.

Cook's tips

This casserole, known as Köthener Schusterpfanne, will taste much better if you use home-made chicken stock rather than stock cubes.

LITTLE SPARROWS

Württenburg has often been called the land of the *Spätzle*. Spätzle (literally little sparrows) are tiny dumplings eaten with practically anything and everything and also on their own. They can be varied by working chopped ham or mushrooms into the dough. Larger versions of Spätzle are made with a filling of meat or spinach and are called *Maultaschen* ('mouth bags').

Creamy rice and fruit pudding

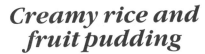

This dish is known in Germany as *Reis Trauttmansdorff*

- **Preparation: 25 minutes**

- **Cooking: 40 minutes**

250g/9oz short-grain rice
1.5L/2½pt milk
75g/3oz sugar
few drops of vanilla essence
300-400g/11-14oz soft or canned
    fruit (see Cook's tips), plus extra
    to decorate
2tsp maraschino (optional)
450ml/¾pt double cream, whipped

- **Serves 4-6**

- **975cals/4095kjs per serving**

1 Wash and drain the rice and set it aside. Pour the milk into a large saucepan, add the sugar and bring to the boil. Immediately add the rice, lower the heat and simmer until the rice is soft, 20-30 minutes. Leave to cool, then stir in the vanilla essence.

2 Meanwhile, prepare the fruit, stoning cherries or cutting any large fruit into raspberry-sized pieces. Put the fruit in a bowl, add the maraschino, if using, and stir well.

3 When the rice is quite cold, fold two-thirds of the whipped cream into the rice. Drain any liquid from the fruit, then fold the drained fruit into the rice mixture.

4 Rinse a 1.5L/2½pt jelly mould with cold water and spoon in the rice mixture. Chill well for at least 1 hour before serving

5 Turn the pudding out onto a serving dish and decorate with fruit and the reserved whipped cream to serve.

Cook's tips

Fresh soft fruit such as strawberries, raspberries and/or cherries are particularly suitable for this recipe, but canned fruit, especially peaches and apricots, may be used provided they have been well drained.

Serving ideas

Accompany this creamy pudding with crisp biscuits. If you don't have a jelly mould, just serve the pudding in an ordinary bowl.

Yeast plum cake

This can be found all over Germany, but its real home is Bavaria. Although the base for the cake is often made with shortcrust pastry, yeast pastry is used here because it soaks up the juice from the plums without becoming soggy

- **Preparation: 45 minutes, plus rising**

- **Cooking: 1 hour**

275g/10oz strong plain flour, plus extra for dusting
1 sachet easy-blend dried yeast
75g/3oz icing sugar
125ml/4fl oz lukewarm milk
50g/2oz melted butter, plus extra for greasing
1 egg
1tsp grated lemon zest
450g/1lb sweet red plums, stoned
sifted icing sugar and cinnamon, for dusting

- **Serves 6-8**

- **315cals/1325kjs per serving**

1 Sift the flour into a large bowl and stir in the yeast and sugar. Make a well in the centre.

2 Beat together the lukewarm milk and melted butter, then beat in the egg and the lemon zest: the temperature should be no more than lukewarm. Pour the egg mixture into the well, then beat with a wooden spoon until the mixture forms a smooth dough.

3 Well butter and flour a cake tin measuring approximately 23cm/9in square – a loose-bottomed tin is best. Then tip out any excess flour.

4 Put the dough in the centre of the tin and press it out with floured hands until it covers the bottom of the tin completely. Halve the plums and cover the dough closely with them, cut sides down and pressed lightly into the dough. Cover the tin with lightly oiled stretch wrap and leave to rise in a warm place, away from draughts, for about 2-3 hours.

5 Heat the oven to 200C/400F/gas 6. When the dough has risen to about twice its original height, remove the stretch wrap, put the tin in the oven and immediately lower the heat to 190C/375F/gas 5. Bake for 45 minutes or until golden

SECRETS FROM THE EAST

East Germany has an almost infinite variety of cakes, mostly made with yeast, of which the *Christstollen*, or Christmas fruit loaf, of Dresden takes pride of place. Families have been known to guard their own special 'secret' recipe for generations, but now, luckily, Dresden *Christstollen* can be bought all over the world.

brown.

6 Dust the cake with a mixture of 1tsp icing sugar and ½tsp ground cinnamon while it is still hot. Serve the cake warm or at room temperature, cut into slices.

Cook's tips

You must use lukewarm milk when making yeast dough in order to activate the yeast – too hot and the yeast will be killed, too cold and the yeast will not be activated.

Variations

Instead of plums, try halved apples. Sprinkle lemon juice over the cut sides to prevent them discolouring.

Traditional Greek Cooking

Greeks eat at a leisurely pace against a background of whitewashed walls, blue sky and fishing boats. Bring this charm to your table with delicious Greek home cooking

Avgolemono soup (page 90)

*T*HE BEST GREEK food is to be found in the village kitchens and the best ingredients are simple ones: the fish freshly caught from the sea, the spring lamb, seasoned with the herbs growing wild on the hillside, cooked over dying embers and served with a salad of tomatoes, peppers and olives. Rice pudding or yoghurt is made special by a spoonful of thyme honey.

Popular produce

Greek cooking is based on the produce of the land. The staple Mediterranean products — corn, grapes, olives and figs — flourish here and, despite the dry and rocky soil, all sorts of vegetables and fruit have been added to the traditional crops. Aubergines, courgettes, tomatoes, onions, beans, artichokes, cauliflowers and pumpkins make the cheap popular dishes, cooked in lavish quantities of olive oil and flavoured with lemon and garlic, dill, fennel or mint.

There are more sheep and goats than there are cattle, as they are content with a meagre diet. Mutton and lamb are the common meats, roasted on the spit or as *souvlakia* speared on skewers and grilled over charcoal. This is one of the most popular ways of cooking in Greece and produces food with a uniquely aromatic flavour. Meat is also cooked in tomato sauce, or very slowly in the oven with vegetables, until it is meltingly tender. Or it can be minced to make delicious meat balls called *keftedes*. *Kokoretsi* consists of alternate pieces of lamb's liver, kidney, sweetbreads and heart grilled in a wrapping of caul which keeps them from drying out. Poultry and pig breeding have been well developed and roast suckling pig is one of the joys of late summer.

Fish from the sea

The sea, too, plays its part in the national economy and provides the favourite food of the islands and the coastal areas. Here you are presented with a choice of red mullet weighed for you in the restaurant before it is cooked in oil or over the fire, fresh tuna fish and a huge variety of Mediterranean fish and seafood. Octopus, mussels and prawns are widely available, while squid makes a most pleasant stew cooked with wine.

The black olives of Kalamata and the soft white goat cheese called *feta* are known everywhere in the world, as are the famous Greek dishes such as *Taramasalata*, the coral cream dip made with grey mullet roes; *Avgolemono*, the egg and lemon soup; and the layers of minced meat and aubergines topped with cheese called *Moussaka*.

Greek food can never be entirely strange, for the standards of taste of the Western world were set in ancient Greece.

The Turks who dominated Greece for nearly 300 years brought a range of foods from the Muslim world, still known by their Arab or Persian names. From the Turks the Greeks learned how to make *pilafi*, a rice dish with tomato sauce; the stuffed vine leaves called *dolmades* and the sweet pastries filled with nuts which can be bought at the numerous pastry shops. From the Turks the Greeks also learned to use the skewer and to like sesame meal and seeds.

Café drinking

At the pavement cafés Greeks often drink *retsina*, beer or *ouzo* and eat *mezze*, little appetizers. These are intended to whet the appetite so they are often highly spiced and flavoured with garlic or sharpened with lemon. Mezze may be cheese, a few olives, a slice of cucumber and a quartered tomato supplemented by pistachio nuts, salted almonds or toasted melon seeds bought from a pedlar, or perhaps *taramasalata*, *dolmades*, *keftedes*, or bits of octopus fried in oil. Sometimes these will be served as a side dish or they can make a meal in themselves, to be ended with coffee, served black, small and sweet, accompanied by a cube of *loucoum* (Turkish delight).

Taramasalata

● **Preparation: 25 minutes**

100g/4oz smoked cod's roe
3-4 slices of stale white bread,
    crusts removed
2tbls grated onion
175ml/6fl oz olive oil
juice of 1-2 lemons
pepper
lemon wedges, parsley and
    chopped olives, to garnish
pitta bread, to serve

● **Serves 6**

● **470cals/1975kjs per serving**

1 Remove the skin from the roe. Place the bread in a bowl of water, then squeeze out the excess. Process the roe, bread and onion in a blender or food processor or mash by hand to a smooth paste.

2 Gradually add the oil and lemon juice alternately, a little at a time, beating constantly or blending until the taramasalata is thick enough to hold its shape. Season with pepper to taste.

3 Serve chilled, garnished with lemon wedges, parsley and chopped olives. Provide pitta bread for dipping.

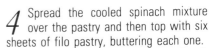

Spinach and cheese pie

- **Preparation: 50 minutes**

- **Cooking: 1 hour**

1 large onion, finely chopped
6 spring onions, finely chopped
4tbls olive oil
900g/2lb spinach
2tbls finely chopped fresh dill or
  fennel leaves, or 2tsp dried dill
225g/8oz feta cheese, mashed with
  a fork
3tbls grated Parmesan cheese
4 eggs, lightly beaten
salt and pepper
pinch of nutmeg
75g/3oz unsalted butter, melted
1 packet (400g/14oz) filo pastry,
  defrosted

- **Serves 6**

- **590cals/2480kjs per serving**

1 Heat the oven to 180C/350F/gas 4. In a large saucepan, fry the onion and spring onions gently in the oil until soft and transparent, but not coloured. Add the spinach with the dill or fennel, cover and cook until tender, turning regularly.

2 Drain the spinach well and leave to cool a little. Add the cheeses and the eggs. Season with salt and pepper to taste and a pinch of nutmeg.

3 Brush the inside of an ovenproof dish measuring about 35.5cm × 25cm/ 14in × 10in with melted butter. Line the dish with a sheet of filo pastry, lifting the edge over the sides. Thoroughly brush the pastry all over with melted butter and repeat with another five sheets of pastry, brushing each one with butter.

FILO PASTRY

Filo, or phyllo, is a paper-thin pastry made in delicate sheets and used in layers for traditional savoury and sweet dishes. It is very difficult to make, so bought pastry is normally used, even in Greece — you can buy it, frozen, in 400g/14oz packs containing about 20 sheets.

Filo will keep, in its wrappings, in the fridge for 3 days or indefinitely in the freezer. To defrost, remove the wrappings carefully and immediately wrap in a dry tea towel, then in a slightly damp one. Leave for 2-3 hours or until completely thawed; it will break if you try to use it while it is still hard.

Peel off only one sheet at a time, re-covering the unused sheets so they don't dry out.

4 Spread the cooled spinach mixture over the pastry and then top with six sheets of filo pastry, buttering each one.

5 Turn the pastry edges under. Brush the top with the remaining melted butter and bake for 45-50 minutes or until the crust is crisp and golden. Serve hot or warm, cut into squares.

Moussaka

- **Preparation: 35 minutes, plus 30 minutes degorging**

- **Cooking: 1 hour 40 minutes**

700g/1½lb aubergines
salt and pepper
2 onions, finely chopped
6tbls olive oil, for frying
700g/1½lb minced lamb
400ml/14oz canned chopped
  tomatoes
2tbls tomato purée
2tsp chopped parsley
1tsp dried oregano
1tsp ground cinnamon
175ml/6fl oz red wine (optional)
For the sauce:
75g/3oz butter
5tbls flour
700ml/1¼pt milk, scalded
pinch of nutmeg
3 eggs, lightly beaten
100g/4oz Parmesan cheese or
  175g/6oz Cheddar cheese,
  grated

- **Serves 6-8**

- **765cals/3215kjs per serving**

1 Cut the aubergines into 5mm/¼in slices, sprinkle generously with salt and leave them to degorge for at least 30 minutes in a colander.

2 Heat the oven to 180C/350F/gas 4. In a large frying pan, fry the onions in 2tbls oil until golden. Add the lamb, breaking it up with a wooden spoon, and cook until brown. Add the tomatoes, tomato purée, herbs, cinnamon, salt and pepper and stir well. Add the wine or 150ml/¼pt water. Simmer for 20 minutes.

3 Rinse the aubergine slices, squeeze a few at a time between the palms of your hands and wipe dry. Fry half of them quickly in half the remaining oil — they absorb less oil if it is very hot. Turn them over once, until they are lightly browned and tender. Drain on absorbent paper. Repeat with the second batch, using the remaining oil.

4 To prepare the sauce, melt the butter in a medium-sized saucepan over low ▶

heat. Add the flour and stir for 3 minutes. Off the heat, add the milk a little at a time, whisking constantly. Season with salt, pepper and nutmeg and cook gently, stirring constantly, until the sauce is thick and smooth. Cool the sauce slightly and stir in the eggs and 3tbls of the grated cheese. 🕐

5 In a 2.3L/4pt baking dish, make three layers of aubergine slices and two layers of meat mixture alternately, starting and ending with aubergines, and sprinkling almost all the rest of the cheese between the layers. Top the dish with the sauce, then the remaining cheese, and bake for about 45-55 minutes or until lightly browned on top.

Plan ahead

The Moussaka can be prepared the day before you plan to serve it. Make the sauce, but do not add the eggs or cheese. Cool completely, cover and chill. The aubergine slices and meat mixture can be layered in the dish with the cheese, then covered and chilled. Before serving, add the eggs and cheese to the sauce and pour over the top aubergine layer, then bake for about 1 hour.

The Greek climate and soil produce palatable wines that are excellent value for money

GREEN CUISINE

Greek cuisine is one of the healthiest in the world. The Greeks eat little meat and rely more on fish, vegetables, pulses and salads. Being a major producer of olive oil, this is used in preference to animal fats. Vegetables are cooked in a very little water — often with olive oil or a sauce — so fewer nutrients are lost. Combinations such as artichokes and broad beans make interesting vegetarian dishes.

Greek desserts can be very sweet, but yoghurt and curd cheese make healthy additions.

Greek honey pie

This pie, from the island of Siphnos, uses a traditional Greek combination of honey and cheese. The soft sheep's milk cheese *mizithra* is used in Greece, but ricotta or an unsalted curd cheese can be substituted

- *Preparation: 30 minutes, plus 30 minutes chilling*

- *Cooking: 45 minutes, plus cooling*

225g/8oz flour
pinch of salt
225g/8oz butter, chilled
450g/1lb ricotta or curd cheese
4 eggs
2tsp ground cinnamon
125-150ml/4-5fl oz clear honey

- *Serves 8*

- *510cals/2140kjs per serving*

1 Heat the oven to 180C/350F/gas 4. To prepare the crust, sift the flour and salt together. Cut the chilled butter into tiny pieces and drop them into the flour. Work the butter into the flour very lightly with your fingertips or a pastry blender. Gradually add just enough water to make the dough hold together in a soft ball

GREEK DRINKS

Greece's most famous wine must be retsina — a white wine heavily flavoured with pine resin. It is something of an acquired taste, so instead you could try the island white, Samos — dry or sweet. Popular reds are Nemea and Mavrodaphne, a rich dessert wine. As an aperitif, have a glass of ouzo, a strong, anise-flavoured spirit. It is usually served in a tall glass, topped up with water.

(about 1-2tbls). Wrap the dough in stretch wrap and chill for 30 minutes.

2 Press the dough into the bottom and sides of a 25cm/10in loose-bottomed flan ring with your hands. Line with greaseproof paper and beans; bake for 15-20 minutes or until dry and crispy, then leave to cool. Remove the beans and paper.

3 Make the filling: beat together the cheese, eggs, 1tsp cinnamon and the honey. Blend well together.

4 Increase the oven temperature to 190C/375F/gas 5. Pour the cheese mixture gently into the pastry case and bake for 35 minutes or until the pie is firm and the top is golden. Dust with the rest of the cinnamon and leave to cool.

Baklava

- *Preparation: 1 hour*

- *Cooking: 1 hour 10 minutes, plus overnight standing*

150g/5oz butter, melted
175g/6oz finely chopped walnuts
175g/6oz finely chopped almonds
100g/4oz caster sugar
1tsp ground cinnamon
2tsp grated lemon zest
400g/14oz filo pastry, defrosted
For the syrup:
350g/12oz sugar
1tsp grated lemon zest
3tbls lemon juice
4 cloves
pinch of ground cinnamon
100ml/3½fl oz clear honey

- *Makes 18-24 pieces*

- *355cals/1490kjs per piece*

1 Heat the oven to 180C/350F/gas 4. Brush the bottom and sides of a 4cm/1½in deep tin measuring about 27cm × 18cm/10¾in × 7in with some of the melted butter. ▶

2 To make the filling, mix the chopped nuts, sugar, cinnamon and lemon zest in a bowl.

3 Cut the sheets of filo in half (you should end with about 24), then trim them to fit the tin. Line the base of the tin with one sheet of pastry and brush with melted butter. Add seven more sheets, brushing each one with melted butter.

4 Sprinkle the pastry with half the nut mixture, smooth the surface then cover with eight more sheets of pastry, brushing with butter as before. Sprinkle on the remaining nut mixture, then cover with the remaining pastry, buttering as before.

5 Brush the top layer of pastry with the remaining butter. Make diagonal cuts through the pastry layers to form a diamond pattern. Bake for 50-60 minutes or until the top is golden brown, covering the baklava with foil if necessary to prevent burning.

6 Meanwhile, make the syrup. Place the sugar, lemon zest and juice, cloves and cinnamon in a heavy saucepan. Stir over a low heat until the sugar has dissolved. Bring to the boil, lower the heat and simmer for about 5 minutes or until pale golden. Remove and discard the cloves. Stir in the honey.

7 Pour the warm syrup over the baklava and allow to stand overnight. 🕐 Before serving, cut through the pastry completely, dividing it into portions. Serve at room temperature.

Plan ahead

Because of its high sugar content, baklava will keep for up to a week. It should be stored at room temperature, covered loosely with foil.

Avgolemono soup

Avgolemono is a sharp-tasting sauce which is used to enrich soups and casseroles. It can also be served as an accompaniment to vegetables such as stuffed vine leaves

- ●*Preparation: 10 minutes*
- ●*Cooking: 30 minutes*

2L/3¹/₂pt chicken stock
100g/4oz long-grain rice
4 large eggs
juice of 2 small lemons
salt and pepper
halved lemon slices, to garnish

- ●*Serves 6* 🍴 ££
- ●*135cals/560kjs per serving*

1 Bring the stock to the boil in a large saucepan, then wash the rice and add to the stock. Return to the boil, then lower the heat and simmer for 15 minutes or until it is tender.

2 In a bowl, mix together the eggs and lemon juice. Stirring constantly, add a little hot stock to the mixture, then pour this into the rice and stock in the pan, stirring. Remove from the heat and add salt and pepper to taste. Serve hot, garnished with halved lemon slices.

Cook's tips

You can use a stock cube for this soup if you must, but home-made stock will give a much better flavour.

Italia, Prego!

Venice and Piemonte show that there is more to Italian food than pasta and pizza

Scampi al terrazzo with plain boiled rice (page 92)

*T*HESE TWO AREAS of Italy rarely use the rich tomato and garlic combinations of the sunnier south and produce specialities based on the proximity of the Alps – Piemonte – and the Adriatic – Venice and the Veneto.

Sturdy staples

You will, of course, find pasta and pizza in both Venice and Piemonte, but they are mainly there for the timid tourist. Local people prefer rice dishes and delicious polenta as an accompaniment to their main course and no-one should leave Venice without treating themselves to a succulent seafood risotto, cooked to order and well worth waiting for. *Risi e bisi*, rice with peas, is another traditional Venetian dish which should be tasted.

Polenta is an old-fashioned 'cake' made from fine cornmeal and baked or fried. It is well suited to the cold mountain climate of Piemonte where it originated, is a favourite in both areas of Italy.

Fishing in the lagoon

It is Venice's long-term relationship with the sea which sets her cooking apart – fresh fish from the lagoon is brought in daily and is sold in the great fish market by the Rialto bridge. *Zuppa di pesce* (fish soup) is a marvellous mix of mussels, prawns, squid, eel and other fishy items, a meal in itself, and local sardines are cooked in a special sweet and sour sauce called *saor* – not to be missed!

Pleasures from Piemonte

In the north, cooking is simple and hearty as befits the rugged Alpine scenery, but when you move towards Turin, a degree of delicacy appears - there will be white truffles on the menu, perhaps, light and melting potato gnocchi and delicate *agnolotti*, the Piemontese version of ravioli. The famous Chicken Marengo (opposite) was invented in this region, a tribute to the exigencies of Napoleon and the ingenuity of his chef. There are sinfully rich goodies to be found in the pastry shops; most tempting of all are the *gianduiotti*, chocolates filled with a soft melting paste of hazelnuts and even more chocolate.

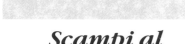

Scampi al terrazzo

This delicious dish is reminiscent of hot, exotic settings – perhaps a terrace overlooking the Adriatic. Complete the scenario with a crisp, dry white wine – a Soave would be the ideal choice.

● **Preparation: 15 minutes**

● **Cooking: 8-10 minutes, plus cooking rice**

1kg/2lb frozen peeled scampi, thawed and drained
2tbls butter
4tbls olive oil
2 garlic cloves, finely chopped
3-4tsp lemon juice
6tbls freshly chopped parsley
2tbls capers, drained
boiled rice to serve
lemon slices to garnish

● **Serves 4**

● **600cals/2520kjs per serving**

1 Pat the scampi dry with absorbent paper. Heat the oil in a large heavy-based frying-pan and cook the scampi very gently over low heat, turning them now

Shopping alongside a canal in sunny Venice

and then, for 6 minutes, until almost cooked. Add the butter.

2 Stir the garlic, lemon juice, parsley and capers into the frying-pan and cook, stirring, for a further 1-2 minutes.

3 Spoon the scampi and cooking juices over a bed of plainly boiled rice arranged on a serving platter. Garnish the edge of the platter with lemon slices. Serve immediately.

Risotto with spring vegetables

Primavera means 'spring', and this classic risotto has been made for centuries in Venetian homes. For a first-class risotto, the vegetables should be tender, new and fresh. Arborio rice, if you can get it, is the best rice for risottos. It absorbs more liquid than other rice, so increase quantities of stock as necessary.

● **Preparation: 25 minutes**

● **Cooking: 40 minutes**

1.2L/2¼pt home-made chicken stock
1 onion, finely chopped
1tbls vegetable oil
75g/3oz butter
250g/9oz Italian rice
1 carrot, finely diced
2 celery stalks, finely diced
100g/4oz shelled peas, or frozen petits pois, thawed
100g/4oz French beans, in 1.5cm/ ½in pieces
1 courgette, finely diced
2 ripe tomatoes, blanched, skinned,

seeds and juice removed, and the flesh diced
75g/3oz Parmesan cheese, freshly grated
salt and freshly ground black pepper

● **Serves 4**

● **510cals/2140kjs per serving**

1 Bring the stock to a simmer in a saucepan. Sauté the onion in the oil and 50g/2oz of the butter in a heavy-bottomed saucepan over low heat until the onion is translucent and golden.

2 Stir the rice into the onion for 1 minute to coat it with butter. Add 200ml/7fl oz of the stock and stir constantly with a wooden spoon, loosening the rice from the sides and bottom of the pan. When the stock has been absorbed, add another 150ml/5fl oz of stock and continue to stir the risotto constantly.

3 When the rice has cooked for 10 minutes, add the carrot, celery, fresh peas, if using, and French beans. Continue stirring and add 150ml/5fl oz of simmering stock when the liquid has been absorbed.

4 After another 10 minutes, add the courgette, the diced tomato flesh and,

1 Wash and dry the chicken pieces and rub them with the cut surface of the lemon, then coat them lightly with the flour.

2 Heat the oil in a large frying-pan over medium-high heat, add the chicken pieces and fry until they are a rich brown on all sides. Reserving the cooking oil, transfer the chicken to a casserole together with half the butter. Pour in the brandy, season with salt and pepper and turn the pieces over, laying them skin side down in the bottom of the casserole.

3 Add the tomatoes, garlic and the stock to the casserole, cover and cook slowly until the chicken is ready, about 40 minutes. Test by pricking the thickest part of the leg: the meat is done when its juice runs out clear.

4 Ten minutes before the chicken is done, melt the rest of the butter in a

---

if using, the thawed frozen peas. Continue to cook, stirring and adding stock carefully as necessary.

5 When the rice has absorbed all the stock, about 30 minutes, remove the pan from the heat, add the remaining butter and half the Parmesan cheese and check the seasoning. Cover the pan for 2 minutes, mix well and turn the risotto into a warmed serving dish. Serve it immediately accompanied by the remaining Parmesan.

Chicken Marengo

Marengo is an area in southern Piemonte where Napoleon won one of his fiercest battles against the Austrians. There his chef created *pollo all Marengo* from the ingredients at hand.

- **Preparation: 30 minutes**

- **Cooking: 1 hour**

*1.6kg/3½lb chicken, cut into
   serving pieces*
1 lemon, quartered
50g/2oz flour
3tbls vegetable oil
40g/1½oz butter
3tbls brandy
*salt and freshly ground black
   pepper*
flat-leaved parsley, to garnish
For the sauce:
*450g/1lb fresh tomatoes, blanched
   and skinned or canned tomatoes,
   roughly chopped*

1 garlic clove, chopped
125ml/4fl oz good meat stock
100g/4oz button mushrooms
*4 slices of good quality white
   bread, cut in half*
2tbls parsley, chopped

- **Serves 4** (♨) (££)

- **670cals/2815kjs per serving**

LUXURIOUS LIVER

Many Italian restaurants in Britain will have *fegato alla Veneziana* on their menu and this simple way of cooking liver makes it a special treat. The secret is in the very thin slicing of the calves' liver and the extra-slow cooking of the accompanying onion. The liver should be still pink and the onions pale, soft and melting. Sage flavours liver well, too.

frying-pan over medium heat and add the mushrooms. Cook them briskly for 5 minutes, stirring, then add them to the chicken.

5 Heat the reserved oil and when very hot, slip in the bread slices. Fry until golden on both sides. Keep them warm.

6 Add the parsley to the chicken and turn the pieces over. Taste to check the seasoning. Transfer the chicken to a heated serving dish. Boil the sauce quickly to reduce if necessary, spoon it over the chicken, garnish and serve.

Coffee and brandy trifle

The name of this pudding, Tiramesu, means "pull me up" in Venetian dialect, because the coffee and brandy have an invigorating effect. Ideally it should be made with a rich Italian cream cheese called *Mascarpone*, but egg custard is a perfectly good substitute.

- *Preparation: 20 minutes, plus 6 hours cooling*
- *Cooking: 30 minutes, including custard*

450g/1lb sponge cake
200ml/7fl oz strong Italian coffee
2tsp coffee essence
100ml/3¹/₂fl oz brandy
For the custard:
3 medium-sized egg yolks
100g/4oz castor sugar
3tbls flour
500ml/18fl oz milk
¹/₂tsp vanilla essence
For the decoration:
300ml/10fl oz thick cream, whipped
1tsp finely ground Italian coffee

- *Serves 8*
- *460cals/1930kjs per serving*

1 To make the custard, put the egg yolks and sugar in a heavy saucepan. Beat the mixture with a wooden spoon until it is pale and creamy and add the flour, 1tbls at a time, beating constantly.

2 In another saucepan bring the milk to simmering-point. Pour the milk on to the egg mixture, stirring all the time. Put the pan over very low heat and cook for 3 minutes or until an occasional bubble breaks through the surface. Remove the pan from the heat, add the vanilla essence and stir well. Leave to cool completely.

3 Cut the cake vertically into 1.5cm/½in thick slices and cover the bottom of a 15 × 20cm/6 × 8in deep cake tin with one layer of cake, patching if necessary.

4 Mix together the coffee, coffee essence and brandy in a small bowl. Brush some of this mixture over the cake slices in the tin. Cover the cake with some of the custard, place another layer of cake slices on top of the custard and repeat the operation until the ingredients are used.

5 Refrigerate the trifle for at least 6 hours. Turn it out on to a serving dish, pipe whipped cream around the base and, at the last minute, sprinkle the top with the ground coffee and pipe with cream.

A Slice of Italy

*I*T IS TUSCANY and its capital city, Florence, which seem to represent the magic of Renaissance Italy and there can be no doubt that this special magic is echoed in some of the traditional dishes of the region. Tuscan cooking is said to have been based on plain but perfect raw materials – the superb bistecca alla florentina, for example. This is a T-bone steak of prime quality grilled, then served with a small amount of olive oil rubbed into both sides. The rich and powerful Medici went on to create culinary masterpieces of great splendour to impress their rivals and their foreign visitors, but at the heart of all Tuscan food is a sturdy peasant simplicity.

Morning markets

Early in the morning, the fruit and vegetable market in Florence is packed with the best produce from the whole of Tuscany: the tiniest artichokes, the red tomatoes, the shiniest aubergines. And of course there are heaps of fresh green spinach for preparing the famous Eggs Florentine, where boiled eggs in a cheese sauce lie on a bed of it.

Alla bolognese

Bologna la Grassa, Fat Bologna, is a city where food rules supreme and meat is the favourite ingredient. The tasty meat and tomato *ragù* which makes the great bolognese sauce for pasta finds its true flavour here in its

Nutty cream fantasy (page 98)

home territory, the local accompaniment for spaghetti or *fettucine alla bolognese*, home-baked lasagne (page 97). The butcher's shop offers a wonderful variety of salami and sausages; the local speciality is mortadella, spicy pork packed with peppercorns and pistachios.

Pleasures from Parma

Pasta without its ritual sprinkling of Parmesan is unthinkable: this hard grating cheese is known locally as *grana* because of its rough grainy texture. Connoisseurs say that it should be left to mature for at least five years and one of the great

95

pleasures of buying it in Italy is tasting your way through Parmesan of different ages. The other great pleasure from Parma is the wonderful Parma ham, *prosciutto,* eaten by itself with crusty bread and butter or accompanied by juicy melon or fresh figs. Thinly sliced and combined with peas and cream, it makes a wonderful sauce for tagliatelle; with Parmesan, mortadella and nutmeg, it forms the basis for ravioli and tortellini stuffings.

A touch of Chianti

Between Siena and Florence lies the wine-growing area known as the Chianti and for many people the wine called Chianti, a dry red wine, is the only wine to drink with Tuscan food. Try one of the many varieties to find the perfect match for the dish you have in mind.

A scene showing the interior of a restaurant in Florence. Fresh produce is stacked up ready for use

Tortellini with butter and cheese

This recipe is for *tortellini asciutti,* which means it is cooked in water and drained. It may be dressed either with butter and Parmesan, with a plain tomato sauce or with cream and Parmesan

- **Preparation: 1 hour**

- **Cooking: 7-8 minutes**

150g/5oz lean pork loin in one piece
150g/5oz turkey or chicken breast, in one piece
40g/1¹/₂oz butter
50g/2oz mortadella, finely chopped
50g/2oz lean ham, finely chopped
2 medium-sized eggs
75g/3oz fresh grated Parmesan
freshly grated nutmeg
salt and freshly ground black pepper
Home-made pasta made with 3 large eggs and 300g/11oz flour
1tbls olive oil
For the dressing:
100g/4oz butter
75g/3oz freshly grated Parmesan
chopped parsley to garnish (optional)

- **Serves 4-6**

- **870cals/3065kjs per serving**

1 Melt the butter in a frying pan over medium-low heat and very gently cook the pork until lightly brown on all sides, about 5 minutes. Add the turkey or chicken breast and cook, turning once, for 3 minutes. Remove the meat and leave it to cool completely.

2 When the meat is cold, chop it by hand as finely as possible. Alternatively chop and mix ingredients in a food processor. In a bowl combine it with the mortadella, ham, eggs, grated cheese and nutmeg. Mix well and season.

3 Make the pasta dough. Use a plain biscuit cutter to cut 24 circles about 4cm/1½in in diameter from the sheet of dough. Cover the rest with a cloth.

4 On each disc put ½tsp of the stuffing. Fold the disc in half over the filling. The top edge should come just short of the bottom edge. Press the edges down firmly to seal. Bring the two points of the semi-circle together, curling it round your index finger into a little ring. Overlap them and press to seal. As you make the tortellini

lay them out on a dry cloth. Repeat the process with 24 more discs. (If you are making pasta by hand-cranked machine, roll out, cut and fill one strip at a time, keeping the rest of the dough wrapped in stretchwrap.)

5 Bring 2.3L/4pt of water in a large saucepan to the boil. Add 2tbls salt and the olive oil to the water. Drop in the tortellini a few at a time and stir gently and continually with a wooden spoon. Fresh tortellini cook very quickly – about 7 minutes. When done, they should be firm but cooked through. Drain and put in a warm serving dish. Stir in the butter and cheese and sprinkle with parsley if wished to add colour.

Variation

Tortellini can also be cooked and served in clear, very delicately flavoured broth.

CROSTINI FOR STARTERS

One of the traditional Tuscan starters, especially in Florence, is the tasty little chicken-livers-on-toast dish, *crostini de fegalini,* served in most restaurants. The chicken livers are sautéed very gently in oil and butter, simmered in wine, then blended with chopped capers, anchovies and garlic and served on small rounds of toast.

Baked green lasagne

A little extra flour will be needed as the spinach makes the dough soft

- **Preparation: 1 hour**
- **Cooking: 30-45 minutes**

Bolognese sauce (fried mince with
  wine and tomato)
butter for greasing and 25g/1oz
  butter
75g/3oz freshly grated Parmesan
For the white sauce:
700ml/1¼pt milk
90g/3½oz butter
75g/3oz flour
freshly grated nutmeg
salt and freshly ground black pepper
For the pasta verdi:
140g/4½oz frozen spinach,
  defrosted
salt
2 large eggs
250g/9oz flour

- **Serves 4-6**
- **1230cals/5165kjs per serving**

1 Make the white sauce, (a butter, flour and milk mix) stir until the sauce is smooth. Season with nutmeg, salt and pepper. Return the saucepan to the heat, and slowly bring the sauce to the boil, stirring constantly. Boil for 2 minutes and remove from the heat. The sauce should have the consistency of thick cream.

2 To make the pasta, cook the spinach with a little salt in a covered pan for 5 minutes. Drain, allow to cool and squeeze out as much water as you can with your hands. Chop the spinach finely.

3 Make the lasagne for home-made pasta, (see page 96) adding the chopped spinach with the eggs to the well of flour in the first step to make it green. Work the flour into the egg and spinach mixture gradually until the mixture has incorporated as much flour as possible, up to 125g/4½oz per egg, without becoming too dry or brittle.

4 Roll out the sheet of dough and cut it into strips 10cm/4in wide. Cut each strips into rectangles 10 × 20cm/4 × 8in. Bring a large shallow saucepan or a roasting tin of water to the boil. Place a large shallow bowl or roasting tin of cold water near the cooker and also spread out some clean, dry tea-towels nearby.

5 When the water boils, add 1tbls salt, drop in 5 or 6 rectangles of dough and stir with a wooden spoon. Cook for 20 seconds after the water has come back to the boil, then lift the pasta out with a slotted spoon and plunge them into the bowl of cold water. Lift them out and spread them on the tea-towels. Repeat until all the pasta has been blanched in this way. Gently pat the pasta dry on top.

6 Heat the oven to 200C/400F/gas 6. Butter a large shallow oven dish and spread 2tbs of the bolognese sauce on bottom. Cover with a layer of lasagne and spread over a little bolognese, some white sauce and sprinkle with grated cheese. Cover with another layer of pasta and repeat until all ingredients are used up. Coat the top layer with white sauce, sprinkle with cheese and dot with butter.

7 Bake the lasagne for 20 minutes or until the top is golden brown, (if necessary grill). Stand for 5 minutes.

Variations

You can use fresh bought lasagne verdi for this dish and increase the parboiling period or alternatively use the non-cook dried variety.

Tuscan chicken with herbs

- **Preparation: 15 minutes, plus 2 hours marinating**
- **Cooking: 45 minutes**

1.6kg/3½lb fresh chicken, jointed,
  or a large guinea fowl, jointed
4tbls flour
salt and freshly ground black
  pepper
25g/1oz butter
4tbls vegetable oil
8-10cm/3-4in sprig of rosemary
4-5 sage leaves
1 sprig of parsley
3 sprigs of thyme
3 sprigs of marjoram
2 garlic leaves
100ml/4fl oz dry white wine

- **Serves 4**
- **550cals/2310kjs per serving**

HUNTING AND SHOOTING

Watch out in the autumn! Woods and meadows are stalked by hunters looking for something for the pot. Pheasants, hares, rabbits and even wild boar are fair game. Roasting is the favourite cooking method and rosemary adds fragrance to the meat.

1 Wash the chicken joints and dry them well with absorbent paper. Season the flour with salt and pepper and coat the chicken lightly with the flour mixture.

2 Heat the butter and oil in a large frying pan over high heat and when very hot, fry the poultry joints, turning until they are a rich brown colour on all sides. Put the joints in a clean plastic bag.

3 Chop all the herbs and garlic together, mix with the wine and pour over the joints. Leave them to stand for at least 2 hours, turning the bag over regularly at least 4 times. Refrigerate at the end of this time if not cooking immediately.

4 Heat the oven to 200C/400F/gas 6. When the oven is hot, pour the wine marinade over the birds, then turn the joints with the skin side downwards. Cover with foil and bake for 45 minutes or until the chicken is tender.

Cook's tips

This dish is excellent served hot with vegetables or cold with salad.

Nutty cream fantasy

The Italian name of this dessert, zuccoto means 'little pumpkin' and refers to its shape, as it is traditionally made in a bowl. A charlotte mould, however, is easier to turn out. It is sold all over Italy, but it is quite easy to make at home

---

● *Preparation: 45 minutes,*
   plus 12 hours chilling

● *Cooking: 15 minutes*

50g/2oz almonds, blanched
50g/2oz hazelnuts in their skins
melted butter for greasing
450g/1lb rectangular, bought
   Madeira cake
2tbls brandy
2tbls Cointreau
2tbls maraschino or other sweet
   cherry liqueur
175g/6oz bitter chocolate, cut into
   small pieces
600ml/1pt thick cream
100g/4oz icing sugar, sifted
For the decoration:
sifted icing sugar for sprinkling,

plus an extra 2tbls
1tbls cocoa, sifted

● *Serves 8* (¶) (££) (⏱)

● *885cals/3715kjs per serving*

---

1 Heat the oven to 200C/400F/gas 6. When hot, put the almonds and the hazelnuts onto two baking sheets and put them in the oven for 5 minutes.

2 Rub off as much of the skin of the hazelnuts as possible and chop them with the almonds. Reserve the nuts. Alternatively use prepared chopped nuts.

3 Grease a 1½L/2½pt bowl thoroughly with melted butter, then line it with stretch wrap, pressing it in place. Generously grease the stretch wrap.

4 Cut the Madeira cake into 1cm/½in thick slices. Moisten the slices with the brandy and liqueurs and line the mould: arrange 3 slices in the bowl from the trim to the centre, cutting them to a point where they fit together. Cut 3 more pieces slightly larger than the remaining gaps, trimming them to fit. Press them tightly in place. Reserve the rest.

5 Melt 50g/2oz chocolate pieces in a bowl set over a small saucepan of barely simmering water.

6 Whip the cream until just stiff and fold in the sugar. Mix the chopped nuts and 100g/-4oz chocolate pieces into the cream. Spoon one half of the cream into the cake-lined mould, spreading it out over the entire lining in a thick layer.

7 Add the melted chocolate to the remaining cream mixture and mix thoroughly. Spoon it into the mould to fill the cavity and level the top.

8 Moisten the reserved cake, if necessary, with brandy and liqueurs. Cover the top of the mould with the smaller pieces of cake, filling any gaps. Cover the mould with cling wrap and weight with a plate. Refrigerate at least 4 hours.

9 Up to 30 minutes before serving, invert a round serving dish over the mould and turn out the dessert. Remove the cling wrap.

10 Cut out a circle of greaseproof paper 26cm/10½in in diameter. Fold the circle in half and then in half twice more, to make eighths. Open the paper and cut out each alternate section, without cutting through the centre where the sections join.

11 Sift icing sugar all over the zuccoto. Mix 2tbls icing sugar with the cocoa. Arrange the cut-out paper pattern over the zuccoto and sprinkle the cocoa and sugar in the cut-out sections. Serve immediately.

When in Rome ...

*you'll find simple, traditional dishes which are easy
to make at home — and they'll bring back all the flavour
of any Roman holiday*

Spaghetti alla carbonara (page 100)

*R*OMAN COOKING HAS changed since the legendary times of lavish meals when gluttonous bouts of eating went on for days. But history has still influenced the traditional dishes of the region.

To the early Romans, the only known method of preserving meat and fish was salting. So, to disguise any staleness or saltiness, preserved food was often cooked with honey, sweet wine or even vinegar. This produced a sweet and sour flavour, known as *agrodolce,* which is still popular today.

Salty meat is also represented by the bacon used in Roman dishes such as Chicken *alla romana,* Roman peas and *Spaghetti alla carbonara* — a pasta dish often cooked as a filling main course in winter.

The area around Rome produces fine-flavoured vegetables, including peas, broad beans and probably the best artichokes in Italy. Herbs are also an important part of Roman cooking. Oregano, rosemary and bay are widely used. Basil, popular throughout Italy, is used as a natural partner to tomatoes, and sage is added to meat and offal to enhance the flavour.

In general, meat dishes tend to come from young animals — chiefly

99

because the poor grazing in the region cannot support large herds.

Sheep's milk is the source of the two most famous local cheeses — pecorino and ricotta. Pecorino is a hard cheese which can be used like Parmesan — grated and added to pasta dishes and soup. Ricotta is a soft cheese made from whey which must be eaten very fresh. It is delicious on its own, sprinkled with salt and pepper.

The Italian sweets and cakes you see in *pasticcerie* and restaurants are often very rich and elaborate. Zabaglione, however, is simple and easy to make, though Romans are just as likely to prescribe it as a pick-me-up as they are to choose it as a dessert.

Naturally enough, Roman wines are the best accompaniment to Roman food. Try Frascati — the dry white wine produced from vines which are grown in the hills of the Lazio countryside.

Spaghetti alla carbonara

Literally translated, this means 'charcoal burner's-style' spaghetti and is served in *trattorias* all over Rome

- **Preparation: 15 minutes**

- **Cooking: 15 minutes**

25g/1oz butter
100g/4oz bacon, cut into short, thin lengths
salt and pepper
400g/14oz spaghetti
2 eggs, beaten
25g/1oz grated Parmesan cheese, plus extra for serving

- **Serves 4** ①£

- **590cals/2480kjs per serving**

1 Melt the butter in a small saucepan, add the bacon and fry gently until cooked, but not crisp.

2 Meanwhile, bring a large pan of salted water to the boil and add the pasta. Return to the boil and cook for about 8-10 minutes or until the pasta is just firm to the bite (*al dente*).

3 While the bacon and pasta are cooking, combine the eggs and cheese in a large, warmed serving bowl. Season to

taste. As soon as it is cooked, add the bacon and the fat from the pan and stir well. Then stir in the drained spaghetti — the heat of the pasta will cook the eggs lightly. Serve at once, with extra cheese handed round separately.

Cook's tips

You can substitute macaroni or tagliatelle equally well for the spaghetti. Whichever type of pasta you use, add 1tbls olive oil to the boiling cooking water to prevent it sticking together.

The warm Italian climate means that evening meals are often eaten outside, or al fresco

ANTIPASTO

Antipasti — which means 'before the pasta' — are served at the start of a meal. One of the most common *antipasto* dishes consists of a variety of thinly sliced cold meats. *Salami* are the best known of these and there will probably be the paper-thin slices of Parma ham. *Mortadella* — a sausage sometimes studded with pistachio — might also be included, as well as *coppa,* made from cured shoulder of pork.

Young vegetables marinated in olive oil and wine vinegar add colour and flavour. Of these, mushrooms, olives, artichoke hearts and pimientos are the most common. You can also add anchovies, canned tuna fish and wedges of hard-boiled egg.

Stuffed artichokes

- **Preparation: 40 minutes**

- **Cooking: 1 hour**

4 globe artichokes
100g/4oz dried breadcrumbs
2 garlic cloves, crushed
2tsp finely chopped mint leaves
1/4tsp freshly grated nutmeg
1/2tsp each salt and pepper
100ml/4fl oz olive oil
2tbls lemon juice
chopped mint leaves, to garnish

- **Serves 4**

- **335cals/1405kjs per serving**

1 Heat the oven to 180C/350F/gas 4. Holding the artichokes upside down, strike them against your work surface to make the leaves open easily. Cut off the stems evenly at the base so the artichokes will stand upright. Remove the tough outer leaves and trim 2.5cm/1in from the tips of the remaining leaves.

2 Mix together the breadcrumbs, garlic, mint, nutmeg, salt, pepper and oil. Pull the outer leaves of each artichoke back and cut away the inner purple leaves.

3 Remove the prickly choke above the base of each artichoke using either a knife or a spoon.

4 Spoon in the breadcrumb mixture and reshape the artichokes. Place in a casserole in one layer and pour round water to a depth of 2.5cm/1in. Cover the pan and place in the oven for 1 hour or until tender when tested with the point of a sharp knife. Serve hot, sprinkled with lemon juice and garnished with mint.

Serving ideas

To eat these artichokes, pull off the leaves one by one and eat the base only, pulling the soft flesh off with your teeth. Next, eat the stuffing and base with a knife and fork.

Provide your guests with a plate for the discarded leaves; fingerbowls are also a good idea.

ARTICHOKE SPECIALITY

You might find several varieties of artichoke in the markets of Rome; some are tender enough to be eaten whole. *Carciofi alla giudea* (Jewish artichokes) are an ancient Roman speciality — the tiniest, tenderest artichokes are cooked whole in a deep pan of very hot oil until golden brown. Unless you grow your own artichokes, you'll probably have to wait until you get to Rome to try them, though, as the artichokes sold in Britain are much larger and less tender.

Roman peas

- **Preparation: 10 minutes**

- **Cooking: 20 minutes**

2tbls olive oil
1 onion, finely chopped
6 bacon rashers, finely chopped
450g/1lb frozen peas
salt and pepper
pinch of sugar

- **Serves 4**   ①£

- **245cals/1030kjs per serving**

1 Heat the oil in a saucepan and gently brown the onion. Add the bacon and cook over medium heat for 10 minutes, stirring occasionally.

2 Add the peas, salt, pepper and sugar, cover and cook over low heat for 10-15 minutes. Serve immediately.

Chicken alla romana

- **Preparation: 20 minutes**

- **Cooking: 45 minutes**

100ml/3½fl oz olive oil
1.4kg/3lb chicken, jointed
1-2 garlic cloves, sliced
50g/2oz bacon, diced
salt and pepper
2tsp chopped fresh rosemary
150ml/¼pt dry white wine
2tbls tomato purée

- **Serves 4-6**   ①££

- **595cals/2500kjs per serving**

1 Heat the oil over medium heat in a casserole large enough to take the chicken joints in a single layer. Cook the garlic and bacon until just crisp, then remove with a slotted spoon and reserve.

2 Blot the chicken joints well with absorbent paper and rub all over with salt and pepper. Add to the pan and cook over medium heat, turning them regularly, until golden all over.

3 Drain off any excess oil. Sprinkle the chicken with rosemary and the reserved bacon and garlic, add the white wine and the tomato purée mixed with 100ml/3½fl oz hot water. Cover and simmer for 25-30 minutes or until tender. Season to taste before serving.

COFFEE TREATS

Almost equal to their love of ices is the Italians' love of coffee, no less so in Rome.

One of the most pleasurable pastimes in this beautiful city is to stroll, eating a *granita* to keep you cool. There are whole shops given over to serving nothing but this delicious coffee water ice, sold for take-away in plastic cups, submerged under a layer of thick cream.

As an after-dinner drink, coffee is usually served *espresso* – very strong and black in very small cups. For morning coffee the familiar *cappuccino* is served, a frothy combination of hot milk and strong coffee topped with cream. The addition of chocolate sprinkled on top is not an Italian custom.

Saltimbocca

The name of this dish translates as 'jump in the mouth' — which is literally what the flavours do

● **Preparation: 20 minutes**

● **Cooking: 25 minutes**

8 veal escalopes (about 450g/1lb)
8 thin slices of ham
8 fresh sage leaves
salt and pepper
25g/1oz butter
100ml/3¹/₂fl oz dry white wine

● **Serves 4**

● **280cals/1175kjs per serving**

1 Beat the escalopes out flat with a rolling pin and trim the ham slices so that they are each the same size as the escalopes.

2 Cover each slice of veal with a slice of ham and a sage leaf. Roll up and secure with a toothpick, 'stitching' the meat so that the toothpick lies along the side of the roll. Season with salt and pepper.

3 In a large frying pan, gently fry the veal rolls in the butter until they are browned all over. Add the wine and simmer for 1 minute, then cover and cook over low heat for 15 minutes or until tender, turning occasionally. Serve the escalopes at once.

Variations

Instead of ordinary cooked ham, you can use Parma ham in the meat rolls for a sweeter flavour to contrast with the veal.

Creamy chocolate cassata

● **Preparation: 45 minutes, plus chilling**

● **Cooking: 10 minutes**

oil, for brushing
75g/3oz plus 2tbls sugar
1tsp lemon juice
25g/1oz peeled almonds
100g/4oz unsalted butter, at room temperature, plus extra for greasing
2 egg yolks
25g/1oz hazelnuts, lightly toasted, skins rubbed off, nuts chopped
25g/1oz plain chocolate, chopped
40g/1¹/₂oz cocoa, sifted
250g/9oz Madeira cake, thinly sliced
75ml/3fl oz rum
For the decoration:
150ml/¹/₄pt double cream, whipped
chocolate curls
hazelnuts

● **Serves 6-8**

● **590cals/2480kjs per serving**

1 To make the praline, line a cold baking tray with foil and brush lightly with oil. In a heavy saucepan, melt 2tbls sugar in 1¹/₂tsp water and the lemon juice over very low heat. When the sugar has dissolved, boil over medium heat, without stirring, until golden. Remove from the heat and stir in the almonds. Pour onto the tray and leave to harden.

2 When the praline is set, carefully peel off the foil. Put between two sheets ▶

of greaseproof paper and break it up finely with a rolling pin.

3 Line an 850ml/1½pt loaf tin with foil, then grease the foil. Cream together 75g/3oz sugar and the butter. Thoroughly beat in the egg yolks, one at a time. Divide between three bowls. Add praline to one, hazelnuts and chocolate to another and cocoa to the third.

4 Place a quarter of the cake slices in the bottom of the tin. Brush the slices with rum and spread them with the praline cream. Cover with a second layer of cake, brush with rum and spread over the chocolate and hazelnut cream. Cover with a third layer of cake, brush with rum and spread with cocoa cream. Cover with cake, brush with rum, cover; chill well.

5 When ready to serve, turn the cassata out onto a chilled serving dish, cover with whipped cream and decorate.

TRUE CASSATA

True cassata is a famous dessert from Sicily, made with cream cheese and an outside layer of sponge cake; it is chilled but not frozen, unlike copies outside Italy which are made with layers of ice cream.

Zabaglione

The Romans make this with dry white wine, though the Sicilian version made with Marsala is better known

- *Preparation: 5 minutes*
- *Cooking: 10 minutes*

6 egg yolks
3tbls caster sugar
175ml/6fl oz dry white wine
crisp biscuits, to serve

- *Serves 4*   (¶¶) (££)
- *215cals/905kjs per serving*

1 Place the egg yolks in the top pan of a double boiler off the heat and whisk with a wire whisk or rotary beater until pale yellow. Gradually add the sugar and beat until the mixture becomes foamy.

2 Place over simmering water and beat in the wine gradually. Continue whisking until the mixture is very thick and triples in size. Pour into four dessert glasses. Serve warm, with crisp biscuits.

Variations

For cold zabaglione, pour the hot dessert into a bowl and place in a larger bowl of ice. Whisk until the mixture is thick and cold, then eat immediately.

Naples Staples

Southern Italy, from Naples down to the Ionian Sea, is famous above all for pizza, tomato sauce and pasta: homely peasant food now enjoyed the world over

CAMPANIA, APULIA AND Basilicata in Southern Italy are among the poorest of its regions. Fruit and vegetables grow in abundance on the volcanic soil, and seafood is plentiful in the fertile coastal strips where most people live. But meat was originally scarce except in the high forest and mountain parts of Apulia and Basilicata, where game, including wild boar, roamed freely. Apulia, down in the 'heel' of Italy, and nicknamed its granary, produces most of the durum wheat used to make pasta and bread.

Long pasta with garlic and oil (page 106)

These conditions produced a diet based on pasta, vegetables, fruit and bread, plus seafood in the coastal areas, and led to one of the great culinary achivements of all time: the pizza. The word means pie in Italian, and the dish goes back to ancient times. It was not a pizza as we know it, for the tomato was missing, as this was unknown in Europe until the early 16th century when the Spaniards brought it back from Mexico.

JUST DESSERTS

As fruit is so plentiful, meals in Southern Italy usually finish with a bowl of fruit, including peaches, grapes and figs. Or there might be a dish of baked peaches, or baked figs stuffed with almonds and fennel seeds. Soft cheese is often used to make desserts, such as Sweet ricotta pies (see recipe) or cheese cakes, also made with ricotta. Cassata, which has become an ice cream outside Italy, was originally made from cream cheese enclosed in sponge cake and chilled.

105

Only then did the pizza napoletana as we know it come about: the disc of crisply baked yeast dough covered with tomato and Mozzarella cheese, and sprinkled with olive oil, oregano and sometimes anchovy fillets. Later it became more than a peasant food: pizza Margherita is named for the first queen of Italy and is based on the national colours: red tomatoes, green oregano and white Mozzarella cheese. This cheese was originally made from buffalo milk, as the arid region did not support dairy cows, but today is usually made with cow's milk.

Pizza 'pie'

To this day pizza is enjoyed in Naples as a convenience food which you can buy and eat on the street. As the basic open pizza was not too convenient the pizza calzone was invented, in which the dough is folded over to enclose the filling. Imaginative cooks created endless variations on the original topping, always using vegetables – tomatoes, capers, olives, onions and garlic – plus salami, seafood and cheese.

Pasta was introduced to Italy by the Sicilians, and it too was once a convenience food. Barrows selling cooked pasta with a choice of sauce were once a common sight on the streets of Naples. Again, cooks' ingenuity came into play, creating over 600 different shapes from the basic pastry dough, and a wide variety of sauces, some designed to suit particular pasta shapes. But all can be eaten al sugo (with meat sauce) or al pomodoro (with tomato sauce). The classic Neapolitan tomato sauce (pomarola) is made with fresh plum tomatoes and plenty of olive oil, flavoured with garlic and basil, and cooked for only 5 minutes. Then there are the seafood sauces, especially those made with vongole (clams) and cozze (mussels.)

Fish dishes are very important, and two which have travelled the world are *Fritto misto di mare* – assorted seafood fried in batter – and *insalata di frutti di mare* (seafood salad). Apulia once belonged to Greece, and local cooking features fish soups and stews of Greek origin. These include oysters, from the Gulf of Taranto.

A baker making and cooking pizzas in Naples

Long pasta with garlic and oil

This is one of the simplest and tastiest of all pasta sauces. It is so quick that it can be prepared in the time it takes for the pasta cooking water to come to the boil. It originated in the slums of Naples

Aubergine and egg pie (page 107)

(page 107)

- **Preparation: 6 minutes**
- **Cooking: 16 minutes**

350g/12oz linguine, spaghetti or
  vermicelli
125ml/4fl oz olive oil
3 garlic cloves, very finely chopped
3tbls chopped parsley
1 dry chilli, crushed
salt

- **Serves 4** ⑪ ££
- **555cals/2330kjs per serving**

1 Bring 4L/1.7pt of water to the boil. Add 3tbls salt. As soon as the water boils rapidly, add all the pasta at once. Do not break the long pasta, but ease it gently as it becomes soft. Stir thoroughly to prevent the pasta from sticking together. Cover the pan, bring the water back to the boil, then remove the lid as soon as the water boils. Stir again, then adjust the heat so that the water boils fast without boiling over.

2 Meanwhile, prepare the sauce. Heat the oil in a large frying pan or a large shallow saucepan over medium heat, add the garlic, parsley and chilli and cook for 2 minutes, stirring constantly. Be careful not to burn the garlic.

3 Test the pasta and drain it as soon as it is al dente – firm to the bite. Strain it in a colander and give 2 or 3 brisk shakes, but be careful not to overdrain or it will become too dry.

4 Add the pasta to the frying pan and cook it over medium-low heat for 2 minutes, stirring: then serve.

Aubergine and egg pie

(Tortino di melanzane e uova sode)
Aubergines are delicious in southern Italy – small, glossy and hard. Since they are always good they are never peeled

- **Preparation: 20 minutes, plus 1 hour standing**

- **Cooking: 45 minutes**

900g/2lb aubergines
salt
vegetable oil for frying
450g/1lb canned tomatoes
24 fresh basil leaves, roughly torn
1tbls oregano
2 garlic cloves, very finely chopped
3 large eggs, hard-boiled and peeled
1tbls finely chopped parsley
freshly ground black pepper
75ml/3fl oz olive oil
50g/2oz freshly grated Parmesan cheese
15g/¹/₂oz butter

- **Serves 4**

- **550cals/2310kjs per serving**

1 Peel the aubergines, unless very young and fresh, then slice them into 5mm/¹/₄in thick rounds. Put the slices in a colander, sprinkle with salt and leave to drain for 1 hour.

2 Rinse the aubergines under cold running water drain and pat dry with absorbent paper.

3 Pour vegetable oil into a large frying pan to a depth of 2.5cm/1in and put over medium heat. When hot, fry the aubergine slices in a single layer, turning until golden on both sides. Transfer each batch onto absorbent paper to drain.

4 Heat the oven to 180C/350F/gas 4. Purée the tomatoes in a food mill or through a sieve. Put into a bowl and mix with the basil, oregano and the garlic.

5 Spread 2tbls of the tomato and herb mixture over the bottom of a shallow ovenproof dish, about 25cm/10in square. Cover with a layer of the aubergines and a few slices of the hard-boiled egg. Sprinkle with some of the tomato and herb mixture, a little parsley and salt and a generous amount of freshly ground black pepper. Pour over 1½tbls of the olive oil and repeat these layers until all the ingredients are used.

6 Sprinkle with the Parmesan cheese and the remaining oil. Dot with the butter and cook in the oven for 20 minutes.

7 Remove the pie from the oven and let it cool slightly, as the dish should be served warm rather than piping hot.

Half moon pizza

(Calzone)

- **Preparation: 25 minutes, plus 1 hour proving**

- **Cooking: 20 minutes**

For the dough:
15g/¹/₂oz fresh yeast, or ¹/₂ sachet easy-blend yeast
¹/₂tsp sugar
200g/7oz flour
1tsp salt
1tbls olive oil
For the stuffing:
50g/2oz Ricotta cheese
75g/3oz Italian salami or prosciutto, cut into 1cm/¹/₂in cubes
100g/4oz Mozzarella cheese, cut in thin strips
3tbls olive oil
salt and freshly ground black pepper
To serve:
Neapolitan tomato sauce (optional)

- **Serves 4**

- **485cals/2035kjs per serving**

1 If you are using fresh yeast, dissolve it in 100ml/4fl oz warm water. If you are using dried yeast, mix it with the flour, sugar, salt and oil. Add the liquid which should be only slightly warm, mix well.

2 Turn onto a floured board. Work with your hands to form a smooth ball. Knead for about 10 minutes, until the dough is smooth and elastic. Transfer the dough to a lightly floured bowl. Cover the bowl with a damp cloth folded double. Leave the bowl in a warm, draught-free place for about 1 hour or until the dough has doubled in bulk.

3 When the dough is ready, heat the oven to 240C/475F/gas 9. Break the Ricotta cheese with a fork, and mix in the salami or prosciutto, Mozzarella cheese and 1 tbls of the oil. Season to taste with salt and freshly ground black pepper.

4 Roll out the dough to a 25-28cm/10-11in round, 6mm/¹/₄in thick. Brush the round with 1tbls of oil. Spoon the stuffing on to one half of the disc, leaving a clear

border about 2.5cm/1in all around. Fold over the disc to form a half moon.

5 Seal the edge with your fingers, then with a fork. Brush all over with the remaining oil. With 2 spatulas, place the calzone on a baking tray. Bake for 15-20 minutes or until the edge is golden.

6 Serve the calzone hot, with Neapolitan tomato sauce if wished.

Sweet ricotta pies

In Naples, pastry is usually made without egg. For a richer version of this recipe, you can make the pies with a French sweet pastry. Serve them hot or cold, with or with cream

● **Preparation: 35 minutes**

● **Cooking: 15 minutes**

200g/7oz flour
75g/3oz caster sugar
½tsp. salt
100g/4oz lard or margarine, diced
3tbls iced water
For the glazing:
1-2 egg yolks
For the filling:
75ml/3fl oz milk
½tsp salt
35g/1¼oz semolina, preferably
   Italian
45g/1¾oz Ricotta cheese
25g/1oz candied fruit, cut in small
   pieces
grated zest of ½ orange
grated zest of ½ lemon
45g/1¾oz caster sugar
1 egg

● *Makes 14-16 pies*
● *180cals/755kjs per serving*

1 To make the pastry, sieve the flour and mix in the sugar and the salt. Add the diced fat and the iced water and work it together using your fingertips. Knead quickly and lightly to a smooth soft dough, wrap in stretch wrap and chill for 1 hour.

2 To make the filling, put the milk in a saucepan with the salt and bring it to the boil. Pour in the semolina in an even stream and stir thoroughly over medium-high heat for 5 minutes. Pour the mixture on to a plate to cool.

3 Place the Ricotta cheese, semolina mixture, candied fruit, grated orange and lemon zest, sugar and egg in a bowl, and mix thoroughly with a spoon.

4 Grease two large baking trays. Roll out the pastry on a floured board to 3mm/⅛in thickness. Using a plain biscuit cutter, 7.5cm/3in diameter, cut out 28-32

discs. Put 1tsp of the filling in the middle of half of the discs and cover with the remaining discs. Seal the edges with a little cold water and press down firmly. Place them on the baking trays and refrigerate for at least ½ hour. Heat the oven to 200C/400F/gas 6.

5 Mix the egg with a fork and brush evenly over the top of each pie. Bake the trays of pies for 10 minutes, or until golden, then reduce the heat to 170C/325F/gas 3, bake for a further 5 minutes. Cool on a wire tray.

MAKING RICOTTA CHEESE

If you cannot find Ricotta cheese in the shops, you can easily make it at home. To make 100g/4oz Ricotta cheese, bring 500ml/18fl oz very creamy milk to the boil with ½tsp salt. Add 2tsp lemon juice and simmer very gently for 15 minutes, stirring frequently. Line a nylon sieve with a linen cloth and pour the milk into it. Tie up the cloth, hang it to drip for about 1 hour. Not only can Ricotta cheese be used in recipes, it is delicious served on its own, sprinkled with a little sugar or fruit syrup.

Calabria Classico

While sun-drenched Calabria is home to a simple, wholesome cuisine based on succulent vegetables, fish and pasta, its neighbour, Sicily, specializes in deliciously wicked cassata and ice cream

Sicilian iced bombe (page 113)

CALABRIA, THE PROVINCE at the 'toe' of Italy, and the island of Sicily are separated only by the narrow Straits of Messina. The sunshine, relentless almost all year round, provides perfect conditions for growing fruit in particular, as well as vegetables. These, together with pasta, form the basis of the local diet, with peasant dishes often being the best.

Italy's poorest province for centuries, Calabria has come into its own as a tourist resort and now,

annually, thousands of tourists sunbathe on its beautiful beaches. As a result, the hotels and restaurants offer cooking similar to that of Naples or Sicily. This, however, has little to do with genuine Calabrian cooking which is simple, robust and nourishing.

Wealth of choice

Vegetables, pasta and pork are the three most important elements in Calabrian cooking inland, while on the coast the most popular fish is

swordfish. Aubergines are the most common vegetables, perhaps because they grow especially well in the dry soil of this region. They were introduced into Europe from Latin America by the Spanish and in Calabria are cooked in innumerable ways.

Almost as common as aubergines are tomatoes, red and green peppers and onions. Most local families traditionally preserve vegetables in oil

A harbour scene, Sciacca, Sicily

and peas, often cooked together in colourful combinations.

Though Naples is generally considered the home of Italian pasta, it really originated in Sicily. One of the most popular ways of cooking pasta locally is with sardines (see recipe) as these small fish are found all around the coast. Among the many varieties of excellent seafood, perhaps the best are swordfish and tuna fish. Hake is caught off the coast of Syracuse and is essential eating for anyone visiting the area.

ADDED EXTRA

seasoned with chilli – a delicious snack eaten with the local crusty bread. Pasta with calabrese is another regional dish.

Pork is Calabria's favourite meat and no part of the animal is wasted. A breakfast dish which is eaten throughout Calabria is pork cooked with pasta, tomatoes and chilli.

Rich legacy

Sicily, Italy's largest province and the Mediterranean's largest island, has always been of great strategic importance to invaders hoping to get a foothold on the mainland of southern Europe. Several of these have left their gastronomic mark and none more so than the Saracens, under whom ice cream was first made.

The conquerors' sweet tooth encouraged the locals to acquire skills making rich desserts, such as sorbet, nougat with honey, crystallized fruit and marzipan; many of these are now famous the world over. The chilled cake, *cassata*, contains ricotta cheese which is a legacy of the period when the Greeks ruled Sicily. They were also the first to plant olive trees, and olive oil remains Sicily's favourite cooking fat.

Fruit and vegetables are traditionally displayed in ranks of magnificent colour on the stalls of street markets. The fruits are mainly

Oregano pasta

Mediterranean citrus fruits, above all lemons. Local vegetables include red and green peppers, artichokes, broccoli, asparagus, aubergines, lettuces, courgettes, beans, tomatoes

Aubergines with mozzarella

- **Preparation: 30 minutes, plus draining**
- **Cooking: 1 hour**

500g/18oz aubergines
salt and pepper
oil, for frying
400g/14oz canned tomatoes
4tbls olive oil
1 garlic clove, crushed
400g/14oz penne or other cut
    tubular pasta
25g/1oz butter, plus extra for
    greasing
75g/3oz mozzarella, very thinly
    sliced
1tbls dried oregano
2tbls dry breadcrumbs
50g/2oz grated Parmesan cheese

- **Serves 6**
- **525cals/2205kjs per serving**

1 Peel and slice the aubergines into 5mm/¼in thick rounds. Put the slices in a colander, sprinkle with salt and leave them to drain for 1 hour. Rinse, then pat dry with absorbent paper.

2 Heat a generous amount of oil in a large frying pan – enough to cover the bottom of the pan. Fry the aubergines in batches in a single layer until golden, turn to fry the other side and then put them on a dish lined with absorbent paper to drain. Repeat with the next batch, adding more oil whenever necessary.

3 Purée the canned tomatoes with their juices in a blender or a food processor. Transfer to a small saucepan. Add the olive oil, garlic and salt and pepper to taste and cook for 10 minutes over medium heat, then reserve.

4 Heat the oven to 200C/400F/gas 6. Cook the pasta in rapidly boiling salted water for 10 minutes or until *al dente*.

5 Meanwhile, butter an ovenproof dish and cover with the aubergine slices. Drain the pasta and toss with a little butter and the tomato sauce.

6 Put a third of the pasta into the dish. Cover with half the mozzarella, half the oregano, a little pepper. Repeat the layers. Cover with the remaining pasta and top with breadcrumbs and Parmesan.

7 Dot with the remaining butter and bake for 10–15 minutes. Remove from the oven and let the dish stand for 5 minutes before serving.

Oregano pasta

- **Preparation: 10 minutes**
- **Cooking: 15 minutes**

75g/3oz stale white bread
400g/14oz bucatini or other
    tubular pasta
125ml/4fl oz olive oil
2 garlic cloves, very thinly sliced
1½ tsp dried oregano
salt and pepper

- **Serves 4**
- **645cals/2710kjs per serving**

1 Remove the crust from the bread and break it into crumbs. Drop the pasta into a large pan of rapidly boiling salted water and cook until it is *al dente* – about 10 minutes.

THE ENGLISH CONNECTION

Marsala, the most famous of all Sicilian wines, was invented almost by accident in the late eighteenth century by John Woodhouse, an Englishman from Liverpool. It is a fortified wine and can be either dry or sweet. It can be drunk chilled as an aperitif or at room temperature as a dessert wine. It is used in various recipes, probably the most well known being zabaglione.

2 Meanwhile heat 75ml/3fl oz of the oil with the garlic in a large frying pan over medium heat for 30 seconds. Do not burn the garlic. Add the oregano, breadcrumbs, salt and a lot of pepper. Mix thoroughly.

3 Drain the pasta, pour in the remaining oil and toss well. Add the pasta to the frying pan and cook for 2 minutes over medium heat, stirring. Serve at once.

Variations

If preferred, omit the oregano and substitute pesto sauce which you can buy ready made in jars.

111

Pasta con le sarde

- **Preparation: 50 minutes**
- **Cooking: 55 minutes**

350g/12oz sardines
250g/9oz fennel bulbs, thinly sliced
salt and pepper
125ml/4fl oz olive oil
1 onion, finely chopped
4 anchovy fillets, chopped
50g/2oz sultanas
25g/1oz pine nuts
1 sachet of saffron powder,
   dissolved in 2tbls warm water
   (optional)
350g/12oz spaghetti
oil, for greasing
2tbls dry breadcrumbs

- **Serves 4**
- **810cals/3400kjs per serving**

1 Remove the heads and tails from the sardines. Open the fish out flat by pressing down on their backs with your thumb, then turn them over and remove their backbones. Wash fish and dry well.

2 Drop the fennel into a small saucepan of boiling salted water and boil for 10 minutes. Drain, reserving the cooking liquid, and cut into very thin, short strips.

3 Heat 100ml/3½fl oz of the oil in a large frying pan over low heat, add the onion and fry until soft and golden. Add the fennel and sardines and cook for 10 minutes, stirring often and adding a few tablespoons of the reserved cooking liquid.

4 Meanwhile, heat the rest of the oil in a small frying pan over low heat. Add the anchovy fillets, heat through, mash and reserve them, keeping warm. Heat the oven to 200C/400F/gas 6.

5 Lift out half the sardines from the pan and reserve them. Add the sultanas, pine nuts, the saffron water if using, and salt and pepper to the pan. Cook for 5 minutes, stirring. Stir in the anchovies, remove from the heat and keep warm.

6 Meanwhile, put the remaining fennel liquid in a large saucepan, adding enough water to make up to 3.5L/6pt, and bring to the boil. Add the spaghetti and cook until it is *al dente*. Drain the pasta and toss in the sardine-anchovy sauce.

7 Grease a deep oven dish with a little oil. Pour half of the pasta into it, cover with the reserved whole sardines and tip in the remaining pasta. Sprinkle with the breadcrumbs and bake for 10-15 minutes. Allow to stand 5 minutes before serving.

'RAGING' AUBERGINE

The aubergine, Calabria's most common vegetable, was called. *mala insana*, raging apple, by the ancients who believed it to be poisonous and cause insanity. Soaking it in salted water was thought to rid it of its unpleasant qualities.

White fish with anchovy sauce

Many Sicilian dishes, like this one, are of Greek origin. A firm-fleshed and easy-to-fillet white fish such as hake is particularly suitable for it

- **Preparation: 25 minutes**

- **Cooking: 30 minutes**

1.3kg/2lb 14oz whole hake, cod or
    other firm-fleshed white fish,
    cleaned, with head on
4tbls olive oil
2 sprigs of rosemary
salt and pepper
1/2 lemon, cut into very thin slices
10 anchovy fillets
1 garlic clove, crushed
boiled potatoes, to serve

- **Serves 4**

- **475cals/1995kjs per serving**

1 Heat the oven to 190C/375F/gas 5. Open the fish from the underside and pull out the backbone. Wash and dry the fish well. Brush a little oil inside the fish with a sprig of rosemary. Season, put in the rosemary and lemon slices and re-form the fish.

2 Heat the remaining oil in a saucepan over medium heat, add the anchovies and garlic and cook for 2 minutes, mashing the anchovy fillets to a paste.

3 Lay the fish on a piece of foil large enough to enclose the fish and pour the anchovy sauce over it. Turn the fish over so that both sides are coated. Sprinkle with a little salt and a good deal of pepper. Wrap the fish in the foil, lay it on a roasting tin and bake for 25 minutes or until it is cooked.

4 Remove the fish carefully from the foil and lay it on a warm dish. Remove the rosemary and the lemon slices from the inside. Pour the sauce left in the foil over it and serve the fish at once, with plain boiled potatoes.

Cook's tips

This size of whole fish might be hard to find; you may need to order it from your fishmonger several days before you need it.

STARTER'S ORDERS

Calabrese, a bright green variety of broccoli, gets its name from Calabria, where it was first grown. It is very tasty and comes in larger clumps than the other types of broccoli.

Sometimes known as 'poor man's asparagus', calabrese can be served as a starter, either hot with melted butter or a white sauce, or cold with an oil and vinegar dressing.

Sicilian iced bombe

- **Preparation: 40 minutes, plus cooling and freezing**

- **Cooking: 10 minutes**

500ml/18fl oz milk
6 egg yolks
150g/5oz caster sugar
grated zest of 2 lemons
300ml/1/2pt double cream
50g/2oz almonds, blanched and
    chopped
50g/2oz crystallized fruit, cut into
    very small pieces
2tbls icing sugar, sifted
1–2 drops of vanilla essence

- **Serves 6**

- **475cals/1995kjs per serving**

1 Bring the water in the bottom of a double boiler to simmering point. In another saucepan, bring the milk to simmering point.

2 Meanwhile, beat the egg yolks and sugar in the top pan of the double boiler, off the heat, with an electric whisk until they are light and pale yellow. Pour the hot, but not boiling, milk slowly onto the yolks and sugar, beating constantly.

3 Put the top pan over the barely simmering water and stir constantly until the mixture thickens and coats the back of a wooden spoon. Do not let the water boil or the mixture will curdle. Add the lemon zest to the custard and remove it from the heat. Allow the custard to cool, stirring frequently.

4 When the custard is cold, pour it into a freezer container, cover and freeze. Remove the custard from the freezer after 30 minutes and whisk it well, then cover and return to the freezer. Repeat this five times.

5 Meanwhile, chill a 1.1L/2pt bombe mould or freezerproof bowl. When the custard is frozen, spoon it into the chilled mould, working quickly. Use a spoon to work the custard smoothly and evenly over the bottom and up the sides, leaving a hole in the centre. Cover the mould and return to the freezer until frozen hard.

6 Whip the cream to soft peaks, then fold in the almonds, chopped crystallized fruit, icing sugar and vanilla essence to taste. Spoon the cream into the centre of the bombe and smooth the top. Cover and return to the freezer for at least 1½ hours.

7 Thirty minutes before serving, dip the mould in hot water for 20 seconds, then turn upside down on a serving dish and stand this in the refrigerator. To serve, cut the bombe into thick wedges, dipping a knife in hot water and wiping clean between cutting each slice.

Variations

Try lining the mould with 350g/12oz crushed sweet biscuits mixed with 50g/2oz sugar and 100g/4oz melted butter.

Cassata siciliana

Genuine cassata has an outside layer of sponge cake and an inner one of cheese; it is served chilled, not frozen, unlike copies outside Italy, which consist of layers of ice cream instead of ricotta cheese

● *Preparation: 50 minutes, plus chilling*

● *Cooking: 5 minutes*

450g/1lb ricotta cheese
175g/6oz caster sugar
200g/7oz crystallized fruit
pinch of ground cinnamon
25g/1oz chocolate, chopped
25g/1oz pistachio nuts, chopped
100ml/3½fl oz maraschino or
*   another sweet liqueur such as*
*   Curaçao or Drambuie*
butter, for greasing
450g/1lb Madeira cake, thinly sliced

For the icing:
350g/12oz icing sugar, sifted
2tbls lemon juice

● *Serves 8-10*   ⓊⓊ ⓔⓔⓔ ⓒ

● *740cals/3110kjs per serving*

1 Press the ricotta cheese through a fine-meshed sieve and set aside. Simmer the caster sugar and 150ml/¼pt water in a saucepan over very low heat until clear but not coloured; do not stir.

2 Meanwhile, cut 150g/5oz of the crystallized fruit into small pieces, reserving the best pieces for decoration.

3 When the sugar has dissolved, pour it over the ricotta and stir hard until the mixture is glossy and smooth. Add the cinnamon, chocolate, chopped fruit, pistachios and half the liqueur; mix thoroughly.

4 Grease and line a deep 20cm/8in round cake tin with greaseproof paper. Line the bottom and the sides of the tin with some of the cake slices. Use the trimmings to fill any gaps. Sprinkle the cake with some of the liqueur.

5 Spoon in the ricotta mixture and cover with a layer of sliced cake. Moisten the top layer with the remaining liqueur. Cover with stretch wrap and chill for at least 6 hours or overnight. ⓒ

6 To make the icing, melt the icing sugar with 50ml/2fl oz water and the lemon juice in a heavy saucepan over low heat. Do not let the icing get too hot. When it evenly coats the back of a spoon, turn the cake out on to a flat plate or cake board and pour the icing over the cake, letting a little of it run down the sides. Smooth the icing on top of the cake with a palette knife.

7 Put the cake back into the refrigerator for at least 15 minutes to allow the icing to set. Decorate with the reserved crystallized fruit. Serve chilled.

ON THE BORDERLINE

Ricotta is an Italian curd cheese made from the whey left over from other cheeses – neither a genuine cheese nor a strictly Sicilian ingredient . . .

Portugal on a Plate

Blessed with miles of coastline, beautiful sunshine and a good rainfall, Portugal is a country with a natural abundance of fresh fish, vegetables and fruit

Caldo verde (page 120)

PORTUGAL LIES ALONG the western side of the Iberian peninsula like a stalwart buffer between Spain and the Atlantic Ocean. Nearly three times as long as it is wide, and running due north to south, Portugal packs an enormous variety of climate, scenery and vegetation into a small area.

The mountainous, richly wooded northern part is watered by numerous rivers and is green and fertile. Small, well-sheltered fields sometimes produce two crops a year of such staples as potatoes and the tall, thick-stemmed local cabbage.

Sheep roam the high fells along the Spanish border. Lower down the hillsides maize, vegetables and fruit grow, particularly vines along the valleys of Tua and the Douro Rivers, home of the famous port wine and of several delicious table wines.

Southwards again, the River Tagus with the elegant capital Lisbon at its mouth cuts diagonally across the centre of the country. At this point the green of the north begins to give way to more southerly characteristics. Olive and almond trees and the cork-oaks, with their beautiful red-brown trunks, take over the landscape, punctuating vast tracts of maize and wheat.

Lisbon fish market

Picturesque fishing villages cling to the rocky coastline to the north and south of Lisbon. You can see fishermen laying out their catches along the quaysides in such places, but nothing rivals the pre-dawn spectacle of the great fish market in Lisbon itself. There thousands of kilos of fish are raucously auctioned off at breakneck speed in semi-darkness. Some are bought by individual vendors, mostly women, who arrange the fish in elaborate patterns in huge flat baskets. Then they go off, with their wares balanced on their heads, to sell as the sun's first rays touch the red rooftops.

Below the Tagus stretch the great plains of the Alentejo where plums and some other fruit are grown. However, the plain is chiefly notable for its scent of wild thyme and coriander and for its herds of pigs.

The allure of the Algarve

The southernmost province of all is the Algarve. Almost Mediterranean in character, it is dotted with pretty resorts and is rich in almond, fig and other fruit trees, flowering shrubs and carefully tended plots of succulent early vegetables.

Like the countryside and seas from which it comes, Portuguese food is immensely varied. Unlikely ingredients, such as meat and shellfish, are blended and flavoured with wine, including Madeira, from the Portuguese island of that name. Meat is chiefly pork, from tender roast suckling pig to the smoked Lamego hams and a dozen varieties of sausage. Lamb or kid is also popular; a traditional festival or wedding dish in the Bairrada region is *chanfana*, lamb slowly casseroled in the dying embers of the bread-oven. Offal is also popular and is used in a variety of ways, among the most famous being *dobrada*, tripe cooked with beans.

CHEESE AND WINE

A few Portuguese cheeses deserve mention, particularly the ewe's milk cheeses of Castel Branco, Azeitão and Serra, and the goat cheese of Tomar and Pombal. Port is the most famous Portuguese wine, followed by the young, light Vinho Verde and the Burgundy-style Dão.

A Portuguese woman baking bread in a traditional oven

Sardine salad

- **Preparation: 30 minutes**
- **Cooking: 15 minutes, plus cooling**

8 sardines (about 900g/2lb), gutted and heads removed
3tbls olive oil, plus extra for greasing
4 tomatoes, thinly sliced
1 small red onion, thinly sliced
2tsp lemon juice
2tsp white wine vinegar
2tsp chopped fresh thyme or ½tsp dried
salt and pepper
lemon wedges, to serve
sprigs of thyme, to garnish (optional)

- **Serves 4-6**
- **505cals/2120kjs per serving**

1 Heat the grill to medium and place the sardines on a lightly oiled rack. Grill for 10-15 minutes or until cooked, turning once. Leave to cool.

2 Skin the sardines and remove the bones. Arrange the tomato and onion slices on four to six individual serving plates.

3 Mix together the oil, lemon juice, vinegar, thyme and salt and pepper to taste. Sprinkle over the sardines and salad and serve with lemon wedges, garnished with sprigs of thyme, if wished.

Salt cod and vegetable omelette

- **Preparation: 35 minutes, plus overnight soaking**
- **Cooking: 50 minutes**

600g/1¼lb dried salt cod
oil, for deep frying
400g/14oz potatoes, cut into matchstick strips
100g/4oz butter
4tbls olive oil
2 onions, chopped
2 large garlic cloves, crushed
salt and pepper
8 eggs, well beaten
2tbls chopped flat-leaved parsley
10 stoned black olives, halved

- **Serves 4**
- **860cals/3610kjs per serving**

1 Soak the cod for 24 hours in cold water, changing the water three or four times.

2 Drain the cod, put it in a saucepan and add fresh water to cover. Bring it to the boil, then reduce the heat and simmer very gently for 15 minutes. Drain and reserve.

3 Heat the oil in a deep-fat fryer to about 180C/350F; at this temperature a 1cm/½in cube of day-old white bread will brown in 60 seconds. Fry the potato sticks until pale golden and cooked through but not crisp. Drain well and spread on absorbent paper.

KEEPING UP THE PRESSURE

A popular method of cooking, particularly in the Algarve, is to use a *cataplana*. Looking like two clam shells, the two rounded halves (made of tin, aluminium or copper) of the cataplana fit together tightly, sealing in the steam and flavour; it works on the same principle as a pressure cooker.

SWIMMING IN FISH

The staple fish in Portugal is cod that has been salted and dried to preserve it. It is so popular it could be called the national food. Rumour has it that there are two dried salt cod recipes for every day in the year – making a staggering 730 different recipes! The seas also yield tuna, hake, swordfish, lobster and other shellfish and fat sardines which are especially delicious when grilled along the Algarve quaysides. Lampreys, salmon and shad are to be found in Portugal's rivers and intriguing combinations of fish feature in the aromatic *caldeiradas*, or fish stews, that are regional specialities.

4 Melt half the butter in half the olive oil in a heavy saucepan over very low heat and, when the foaming has subsided, add the onions. Stir, cover and let the onions soften over low heat for 10-15 minutes, stirring occasionally.

5 Meanwhile, using a knife and fork and being very careful, remove the skin and bones from the cod. Flake the flesh into even-sized pieces and reserve.

6 Add the garlic to the onion mixture, raise the heat and stir until golden brown. Add the reserved fish and continue stirring for 7-8 minutes. Remove the pan from the heat, combine the mixture with the cooked potatoes, trying not to break up the fish too much, and season with salt and pepper to taste.

7 Take a large, heavy frying pan or two smaller frying pans and heat the rest of the oil and butter over medium heat. Stir in the fish and potato mixture and, when heated through, add the beaten eggs.

8 Keep the mixture moving slightly with a spatula, tilting the pan to let the egg run evenly and set. When the eggs are set but not dry, slide the omelette onto a warmed serving plate, sprinkle with parsley and olive halves and serve at once.

Piquant casseroled chicken

Long, slow cooking at a very low heat is the secret behind this wonderfully succulent dish. The alcohol in the sauce also helps to make the chicken particularly tasty and tender

● *Preparation: 45 minutes*

● *Cooking: 7½ hours*

1kg/2¼lb chicken, cut into serving pieces
4 thin slices smoked ham, about 125g/4½oz
12 small onions, peeled
1 red pepper, seeded and sliced
4 tomatoes, quartered and seeded
2tbls Dijon mustard
3 garlic cloves, crushed
100ml/3½fl oz port
100ml/3½fl oz brandy
200ml/7fl oz white wine
salt and pepper
flat-leaved parsley, to garnish
boiled rice, to serve

● *Serves 4*   ① ££

● *640cals/2690kjs per serving*

1 Heat the oven to 200C/400F/gas 6. Lay the chicken portions on the sliced ham in a deep earthenware casserole and surround them with the vegetables. Dot the mustard and crushed garlic over the chicken, pour in the liquids and season with salt and pepper to taste. Cover and put the casserole in the top of the oven for 30 minutes.

2 Turn the temperature down to the very lowest setting and cook for a further 6-7 hours. The chicken should be very tender. Check the seasoning, garnish with parsley and serve hot, with boiled rice.

Variations

As an alternative to the boiled rice, try serving this casserole dish with fresh pasta or boiled potatoes.

NO-OVEN COOKING

An ordinary Portuguese household may well not have an oven, there being no call for one: meat is usually boiled, not roasted; soup is often prepared to be a meal in itself; and stews are often strained and pasta cooked in the strained broth, then served after the stew. You certainly wouldn't need the extra heat from an oven, as the Portuguese climate is very warm.

Walnut pudding

- *Preparation: 30 minutes*

- *Cooking: 50 minutes*

40g/1¹/₂oz butter, plus extra for greasing
175g/6oz walnut pieces
100g/4oz caster sugar
3 eggs, separated
¹/₄tsp ground mixed spice
2tbls Madeira or cream sherry
whipped or pouring cream, to serve

- *Serves 4*

- *565cals/2375kjs per serving*

1 Butter a deep 18cm/7in round dish and line the base with a circle of buttered greaseproof paper. Process 100g/4oz of the walnut pieces very finely.

2 Whisk 50g/2oz of the sugar with the egg yolks until very thick and creamy. In a large, clean, dry bowl, whisk the egg whites until stiff, then whisk in the remaining sugar.

3 Carefully fold the processed walnuts, spice and egg yolks into the whites. Pour into the prepared dish. Cover with pleated foil or greaseproof paper and tie in place with string. Place a trivet in a large saucepan and stand the dish on it. Pour in boiling water to come halfway up the sides of the dish.

4 Cover the saucepan tightly and steam the pudding for 30-40 minutes or until it is firm and a knife inserted in the centre comes out clean. Turn out onto a serving dish.

5 Cook the reserved walnuts in the butter for 2-3 minutes, stirring constantly. Add the Madeira or sherry and cook, stirring, for 1-2 minutes more to make the sauce. Allow to cool slightly, then arrange the walnuts on top of the pudding, pour over the sauce and serve with cream.

Cook's tips

If you do not have a food processor, put the walnut pieces in a greaseproof paper bag, secure the opening and crush them finely with a rolling pin.

Freezer

Shelled walnuts, whole or in pieces, will freeze for up to 1 year. Wrap in foil or pack in small cartons. To use, thaw at room temperature for 3 hours.

THE GREAT POTATO RACE

When you eat out in a restaurant in Portugal, whether for lunch or dinner, you will come across an interesting custom: if you order fish it will come with boiled potatoes; if you order meat it will invariably come with chips.

SO MANY SOUPS

Soups, Portuguese-style, are made of a whole range of vegetables, chicken, fish, game and even bread and garlic. Perhaps the most famous soup is *Caldo verde*, a green cabbage soup from northern Portugal (see recipe).

Caldo verde

Translated into English, the name of this dish is 'green cabbage soup'

- **Preparation: 30 minutes**

- **Cooking: 40 minutes**

375ml/13fl oz chicken stock
450g/1lb potatoes, thinly
  sliced
1 onion, finely chopped
300g/11oz green cabbage leaves
25g/1oz butter
salt and pepper
croûtons, to serve

- **Serves 4**   ① ⑤

- **265cals/1115kjs per serving**

1 Make the chicken stock up to 700ml/ 1¼pt with water and bring to the boil in a large saucepan, then add the potatoes and onion. Cover, lower the heat and simmer for 15-20 minutes or until the vegetables are very soft. Remove the pan from the heat.

2 Meanwhile, roll the cabbage leaves tightly together and shred them as finely as possible with a sharp knife.

3 Purée the potato mixture in a food processor or blender, adding the butter and, if necessary, a little water; the soup should be the consistency of thin cream. Season with salt and pepper to taste. Return the soup to the pan and bring it to the boil.

4 Toss in the shredded cabbage and boil, uncovered, for 3-5 minutes, until the cabbage is cooked but still slightly crisp. Adjust the seasoning and serve the soup sprinkled with croûtons.

Sweet rice

- **Preparation: 15 minutes, plus chilling**

- **Cooking: 50 minutes**

850ml/1½pt milk
175g/6oz short-grain pudding rice
75g/3oz sugar
pared zest of 1 large lemon
2 eggs
4 egg yolks
quartered orange slices, to decorate
For the topping:
2-3tsp caster sugar
¼tsp ground cinnamon

- **Serves 4**   ⑪ ££ ⑤

- **515cals/2165kjs per serving**

1 Place the milk, rice, sugar and pieces of lemon zest in a large saucepan. Bring to the boil, then lower the heat. Cover tightly and simmer for 35-40 minutes or until the rice is tender and most of the milk has been absorbed, stirring occasionally. Remove from the heat.

2 Remove the pieces of lemon zest from the rice mixture. In a small bowl, beat together the eggs and the egg yolks, then beat into the rice. Return to a very low heat and cook for 2 minutes, stirring constantly.

3 Place the rice in four individual serving bowls, allow to cool, then chill for 3-4 hours. ⑤

4 To serve, mix together the sugar and cinnamon and sprinkle over each bowl, in parallel lines. Decorate with quartered orange slices and serve immediately, while still chilled.

Cook's tips

Before the zest is pared from the lemon, make sure that the fruit is first soaked in warm water for about 5 minutes and then dried thoroughly with absorbent paper. This will help to remove any coating.

Scandinavian Scoff

Scandinavian dishes, featuring cheese, bacon, ham, herrings and seafood, with unusual soups and sweets, make a unique contribution to European cuisine

S*CANDINAVIAN COOKING HAS* evolved quite differently from mainline European *cuisine*, and if you are seeking an interesting and mouth-watering change you could do much worse than head north – if only in spirit from your kitchen.

All four Scandinavian countries have their own superb national and regional cuisine, some of which has come to be appreciated far outside Scandinavia. Perhaps the best-known is the Swedish *smörgasbord*, a lavish buffet featuring a variety of open sandwiches, herrings treated in several ways, salmon, venison, seafood, hot potatoes and a great many other edibles.

'Burning wine'

Among the drinks at a *smörgasbord* party one invariably finds the fearsome clear spirit, aquavit – Scandinavians call it 'burning wine' – and light lager-type beers.

Norway, with its crystal clear, unpolluted lakes and coastal waters, is a Mecca for fish and seafood connoisseurs. Lobsters, crabs, shrimps and prawns are abundant there, as are trout, cod, plaice, turbot, salmon, eels and plump herrings. The fish market at Bergen is world-famous for the range and variety of its mouth-watering displays.

The Norwegians have their very own ways with fish – for example

Prawn prelude (page 123)

dill-marinated salmon served in thin slices and eaten with a dill-and-mustard sauce and crisp-fried bits of salmon skin; crayfish boiled with dill and strong beer; fresh poached salmon with horseradish mayonnaise; fried trout with a soured cream sauce, served separately.

Reindeer salami

Norway has its own approach to meat too, for example reindeer salami, mutton salami and smoked lamb.

There's more to Denmark than its justly famous cheese, butter, hams and bacon. Like the Swedes, the

Danes are fond of big elaborate open sandwiches, often for lunch, which they call *smoerrebroed* and which are sometimes served with hot dishes. At dinner one might be served the popular yellow lentil soup with pancakes and a main dish of fish with parsley sauce, or of meat in a rich brown sauce with a sweet cucumber salad.

Danes are very fond of their bread and cakes. There is a great variety of light and dark breads.

Seasonal delights

Among the many seasonal Danish delights are February's early lamb, Marsh sole, the white asparagus of April, the lobsters and blue mussels that appear in May and the June crabs and little Limfjords shrimps.

August sees plump crayfish in the shops, and in September there are superb salmon, oysters, eels and cod. September is also the start of the wild duck and deer season.

In Finland the cuisine is based on wholesome fish, game, reindeer and cereals. The dishes are simple and appetizing, and vary with the passing seasons. In wintertime the Finns enjoy hearty soups, stews and casseroles, some of them made with fish. Winter is also the strange spawning time of the fish called burbot, and its roe – called 'Finnish caviar' – is enthusiastically eaten on toast with thick soured cream.

Ongoing feast

In the spring there are lamb dishes, as well as marinated herrings, perch, pike and bream.

Summer is the high time of the great ongoing Finnish feast, with all kinds of shellfish, meats and fruits.

Autumn in Finland is a time of fresh-caught salmon, lavaret and vendace, of grilled lampreys and game dishes.

Finnish regional delicacies are quite unique. In northerly Lapland reindeer steaks and smoked reindeer tongues are much relished, as are wild Arctic strawberries. In the south of the country there are to be found quite delicious indigenous cheesecakes and local dishes based on herrings and eels.

A vegetable stall in Helsinki showing all the available produce

FINNISH FUNGI

Mushrooms are picked throughout Finland, which boasts a staggering 500 different varieties. About 30 of these types are particularly prized for their distinctive flavours and are used generously in a number of traditional meat and game dishes.

Open sandwiches

These sandwiches, literally translated as 'butter bread', are a Danish national dish and almost a meal in themselves. Easy, delicious and very colourful, there are over 100 different types. The Danes can choose from a wide variety of different breads to use as bases for the sandwiches.

- **Preparation: 10 minutes**

For an egg and tomato sandwich:
1 slice of rye bread
softened butter
1 lettuce leaf
1 hard-boiled egg, sliced across evenly
½ firm tomato, cut into wedges
thick mayonnaise
2tsp red caviar
1 sprig watercress
For a shrimp sandwich
1 slice of white bread
softened butter
1 lettuce leaf
50g/2oz boiled, peeled shrimps
3 thin slices of lemon
2 slices of tomato
2 boiled unpeeled prawns
cress or fresh dill

- **Serves 1-2**
- **390cals/1640kjs per serving**

1 For the egg and tomato sandwich, butter the rye bread generously and put the lettuce leaf on top. Lay overlapping slices of egg down one side of the lettuce. Overlap wedges of tomato on the other half. Pipe a little mayonnaise down the centre and garnish with caviar and watercress or mustard and cress.

2 For the shrimp sandwich, generously butter the slice of white bread and put the lettuce leaf on top. Pile the shrimps on the lettuce and garnish with the lemon slices, tomato slices, prawns and cress or fresh dill.

Prawn prelude

Shellfish from Norway

- **Preparation: 15 minutes**
- **Cooking: 5 minutes**

450g/1lb raw unshelled prawns
To serve:
275ml/10fl oz Mayonnaise
lemon wedges
2-3 sprigs fresh dill
12 slices milk loaf

butter
crushed ice

- **Serves 4**
- **650cals/2730kjs per serving**

1 Fill a large pan with cold water, add a pinch of salt and bring to the boil. When boiling drop in the prawns and cook until they are just pink, 3-4 minutes. Cool in the cooking liquid. When cool drain the prawns and dry them.

2 Put the crushed ice on a serving dish and arrange the prawns in circles on the bed of ice. Garnish with sprigs of fresh dill and lemon wedges. Put the mayonnaise in a small bowl. Alternatively, hook the prawns around the edge of a glass filled with crushed ice.

3 Toast the slices of milk loaf and put in a napkin to keep warm. Serve immediately with the prawns, mayonnaise and butter.

Danish delight

This much-loved dish is inexpensive, easy to prepare and an excellent way of using left-over beef, lamb, pork or bacon

- **Preparation: 30 minutes**
- **Cooking: 30 minutes**

450g/1lb cooked, cold lean meat
2 large onions
6 large potatoes, boiled in their skins until tender and allowed to cool
50g/2oz butter
salt and finely ground black pepper
Worcestershire sauce
6 eggs

- **Serves 6**
- **440cals/1850kjs per serving**

1 Cut the meat into 1cm/½in cubes and reserve. Cut the onions and potatoes into pieces the same size as the meat and reserve both separately.

2 Melt one-third of the butter in a large frying pan over medium heat. When it is hot, sauté the meat for 4 minutes, stirring occasionally. Remove the meat from the pan and keep it warm. Repeat with the onions, then the potatoes.

3 Mix the meat, onions and potatoes together in a frying pan and cook for a further 5 minutes, stirring occasionally. Season to taste with salt, pepper and Worcestershire sauce. Pile the mixture onto a heated serving dish and reserve.

4 Meanwhile fry the eggs. Put the fried eggs on top of the sauté mixture and serve piping hot.

Variation

For a more elaborate presentation, serve the dish with béarnaise sauce, cold cooked beetroot, and garnished with 2 hard-boiled eggs in place of the fried eggs.

VILLAGE BAKERY

The Danes are very particular about the quality of their bread, and there are many varieties, light and dark, sweet and savoury. Danish bakers make their own bread, cakes and pastries on their premises, and even in small villages there are often two or more local bakeries.

Karelian stew

A traditional hearty winter casserole dish from Finland, usually eaten with barley bread to keep out the cold

- **Preparation: 30 minutes**

- **Cooking: 3 hours**

450g/1lb good quality stewing
   beef, cut into 4cm/1¹/₂in cubes
250g/9oz lean shoulder of lamb,
   cut into 4cm/1¹/₂in cubes
250g/9oz pork boneless spare rib
   joint, cut into 4cm/1¹/₂in cubes
1 veal kidney, cut into 1.5cm/
   1¹/₂in slices
250g/9oz ox liver, cut into 1.5cm/
   ¹/₂in slices
2 onions, chopped
1tbls salt
10 allspice berries

- **Serves 6**

- **400cals/1680kjs per serving**

1 Heat the oven to 240C/475F/gas 9. Put the beef into the bottom of a large casserole, sprinkle some of the onions, salt and allspice berries on top then repeat with layers of lamb, kidney and liver. Put the pork on top and add just enough water to cover.

2 Put the stew in the oven, uncovered, until the pork lightly browns, about 30 minutes. Lower the temperature to 180C/350F/gas 4, cover tightly and cook for between 2-2½ hours.

Cook's tips

This stew makes an excellent pie filling. Serve with squares of cooked puff pastry.

Cauliflower salad

Lettuce is an expensive hot-house or imported commodity in Norway so cauliflower is often used instead

- **Preparation: 20 minutes**

1 cauliflower
¹/₂ cucumber
1tbls chopped onion
2 tomatoes, cut into wedges
2tbls chopped chives
For the dressing:
6tbls soured cream
2tbls mild mustard
1tbls lemon juice
1¹/₂tsp sugar
salt
freshly ground black pepper

- **Serves 6**

- **50cals/210kjs per serving**

1 Wash the cauliflower, separate the florets and cut into small pieces. Use some of the stalks provided they are not too coarse. Wash the cucumber but do not peel. Cut it into 1.5cm/½in cubes.

2 Put the cauliflower, cucumber and chopped onion into a large salad bowl in layers.

3 To make the dressing, combine the soured cream, mustard, lemon juice and sugar in a small bowl and mix thoroughly. Pour into the salad bowl and toss so that everything is well coated. Season to taste.

4 Garnish with the tomato wedges and sprinkle the chives on top. Serve with boiled sausages, liver, meat patties or fried fish.

Pork patties

This is a traditional dish often served at Christmas in Norway. Boil cabbage in meat stock to accompany the dish and scatter it with caraway seeds. Serve with cooked prunes, or apple sauce

- **Preparation: 20 minutes**

- **Cooking: 15 minutes**

450g/1lb lean pork
225g/8oz veal
1¹/₂-2tsp salt
¹/₂tsp ground black pepper
¹/₂tsp ground ginger
250ml/9fl oz milk
275ml/10fl oz stock
50g/2oz butter or margarine
15g/¹/₂oz flour
flat-leaved parsley to garnish
   (optional)

- **Serves 8**

- **255cals/1070kjs per serving**

1 Put the pork through the coarse blade of a mincer twice. Add the veal to the minced pork and mince 2 more times. ▶

2 Put the forcemeat in a large bowl and add the salt, pepper and ginger. Stir well. Add the milk gradually, stirring all the time.

3 Using a tablespoonful of forcemeat at a time, shape into round, flat cakes. Melt 25g/1oz butter in a heavy frying pan and fry the patties until browned on both sides – about 5-10 minutes each side.

4 Place the meat patties in a heavy saucepan. Add 275ml/10fl oz stock to the frying pan, stir once or twice and then pour this gravy over the meat patties.

5 Melt the remaining butter in the frying pan, stir in the flour and cook to make a brown roux. Stir in the gravy from the saucepan containing the meat patties. Stir constantly until sauce thickens, then pour it over the meat patties and simmer for 5 minutes. Serve garnished with flat-leaved parsley if wished

SMORGASBORD

The origins of the sumptuous Swedish *smörgasbord* are in some dispute, but the nicest theory is that it started as a kind of free-wheeling rural bottle party, all of the guests bringing their own drink and those dishes they could do best. It is an easy form of party-giving for it allows host or hostess to be with the guests throughout the meal.

Loin of pork larded with prunes

A fruity Swedish savoury

● *Preparation: 15 minutes*

● *Cooking: 1 hour*

800g/1lb 12oz boneless loin of pork

12 stoned prunes
2tsp salt
1tsp ground ginger
¼tsp freshly ground black pepper
For the sauce:
1 apple, cut in pieces
1 onion, cut in pieces
4 prunes, cut in pieces
2tbls soy sauce
1tbls flour

● *Serves 4*   (♟) (££)

● *615cals/2585kjs per serving*

1 Heat the oven to 225C/425F/gas 7. Make a hole right through the centre of the loin lengthways with the aid of a wooden spoon handle or similar object. Stuff in the prunes one at a time so that they are uniformly distributed throughout.

2 Tie one end of the loin firmly with kitchen string, then tie it several times along the length at intervals and one long string round the entire length.

3 Mix the salt, ginger and pepper and rub them into the meat. Insert a meat thermometer into the meat if you have one. Put the loin in a roasting tin surrounded by the fruit and onion pieces. Roast the pork in the oven until the thermometer registers 85C/180F (1 hour). Strain pan juices, add soy sauce and heat gently. Thicken with flour and serve the sauce with the pork.

'FORK BITS'

Norwegian supermarkets must be the only ones in the world where you can buy cans of little herring slivers in sweet-and-sour sauces made with lobster, prawns, sherry or tomatoes. They are popular snacks which the Norwegians call *gaffelbiter* which means 'fork bits'.

Swedish coffee bread

● *Preparation: 2 hours*

● *Cooking: 20 minutes*

400g/14oz flour
1 sachet easy-blend dry yeast
¾tsp salt
50g/2oz caster sugar
250ml/9fl oz milk and water
  mixed in equal proportions,
  lukewarm
50ml/2oz butter or margarine,
  melted
1tsp powdered cardamom
1 egg, beaten, for brushing
1tbls icing sugar, for sprinkling
For the filling:

50g/2oz butter or margarine,
  softened
75ml/3oz caster sugar
1tbls caster sugar mixed with
  2tsp vanilla essence or 2tbls
  powdered cinnamon

● *Serves 8-10*

● *365cals/1535kjs per serving*

1 Sift the flour, yeast, salt, cardamom and the sugar into a bowl. Add the liquid and melted fat, mix to a dough. Turn out the dough on a lightly floured surface and knead it until the dough is smooth, cohesive and has a shiny surface.

2 Return the dough to the bowl, sprinkle with a little flour on top, cover with a cloth and leave it in a warm place until it doubles in size, about 1¾-½ hours.

3 Work the vanilla (or cinnamon) sugar mixture with the caster sugar until evenly mixed, and reserve.

4 Work the dough in the bowl for 1 minute, turn it out on a lightly floured surface and knead lightly until smooth, 1-2 minutes.

5 Roll out the dough into a 45cm/18in square, spread the softened margarine over the dough and sprinkle with the filling mixture. Cut the square in half and roll up the two pieces lengthways.

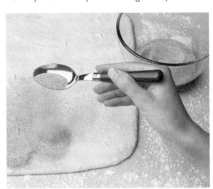

6 Heat the oven to 200C/400F/gas 6. Place the rolls on a lightly greased baking tray. Using scissors, make incisions at regular intervals along the rolls but without cutting right through. Pull the slices to alternate sides, squeezing each top to a point.

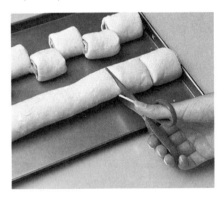

7 Brush the rolled dough with the beaten egg and bake for 15-20 minutes or until lightly golden. Cool for 10 minutes and sprinkle with icing sugar before serving.

Variations

For buns, cut the rolled, buttered and sugared dough through into equal slices. Put these with one cut surface downwards on the greased baking tray either directly or into paper cases. Leave to rise in a warm place for 20 minutes, then bake in the oven preheated to 250C/500F/gas 10 for 10 minutes or until lightly golden.

Viva España

Spanish cooking is robust, flavoursome and lacking in pretension. From rich stews and shellfish in the north to paella and gazpacho in the south, there's something to tempt everyone

*N*ORTHERN SPAIN IS a huge area encompassing several different races and cultures from Galicia and Asturias, the Basque country, Navarre and Aragon, Castile and Catalonia.

In Galicia, fish and shellfish are very popular, whereas in neighbouring Asturias the smoked black blood sausage *morcilla* is made and used to flavour *fabada asturiana*, a stew of white beans and pork.

Basque cuisine

In the north-west corner of Spain, close to the Pyrenees, lie the Spanish Basque provinces. Basque cooks make good use of the tremendous amount of fish and shellfish which are found in the cold waters of the Bay of Biscay, and of the butter and cream from local dairy herds.

Navarre and Aragon are the next two regions eastwards to the south of the Pyrenees. Here the long-

Iced Andalusian gazpacho is a refreshing soup for hot days

haired Lacha sheep which graze on the foothills provide some excellent cheeses as well as lamb.

In Castile, the most famous dish is *Cocido madrileño* or Madrid stew, which has evolved over many centuries. It is a one-pot meal containing pulses, meat, sausages and vegetables. The full-bodied claret-type wines of the Rioja wine-growing district go well with the food as do the wines of La Mancha which include Valdepeñas.

In Catalonia and the nearby Balearic islands, the dishes are more delicately flavoured and include *zarzuela de pescados* – shellfish cooked in a wine and brandy sauce.

Moorish influence
Travelling south from Madrid you will come to the regions of southern Spain: Andalucia, Extremadura, Murcia and Valencia. The influence of 800 years of Moorish occupation is very noticeable here. The use of spices and herbs such as cinnamon, saffron, cumin, mint and coriander in fish and meat dishes, and the very rich eggy, almondy sweets are typically Arabic, as is the addition of orange or lemon juice and ground almonds to sauces.

Andalucian attractions
Andalucia is the birthplace of Gazpacho. It is a dish with very ancient origins and it has many variations. This region is also famous for its olive oil production and sherry, which is made around the town of Jerez de la Frontera.

Along the Portuguese border is the region of Extremadura. Pork and

A satisfying snack, deep-fried churros *are always popular*

game are features of the menu here and these include *chorizo* and *morcilla* sausages. As Extremadura has no coastline, dried salt fish (known as *bacalao*) is also very popular.

Paella perfect
The regions of Murcia and Valencia are famed for saffron and the world-famous paella. Originally a peasant invention using local ingredients, paella was traditionally cooked in a two-handled pan known as a *paellera* on the open fire.

Andalusian gazpacho

● *Preparation: 40 minutes*

2 garlic cloves, chopped
1 green pepper, seeded and chopped
1 cucumber, seeded, peeled and
    chopped
8 tomatoes, skinned and seeded
1tbls chopped parsley
1tbls chopped mint (optional)
12 blanched almonds (optional)
50g/2oz dry white breadcrumbs
2tbls wine vinegar
2tbls olive oil
1tbls tomato purée
salt and pepper
300ml/½pt iced water
ice cubes

For the garnish:
2 hard-boiled eggs, chopped
1 onion, diced
1 red pepper, seeded, finely diced
½ cucumber, diced

● *Serves 4*

● *190cals/800kjs per serving*

1 Put all the ingredients, except the iced water and ice cubes, into a blender or food processor and blend for 30 seconds or until smooth, seasoning to taste with salt and pepper.

2 Turn the mixture into a large serving bowl and add the iced water, cover and place in the refrigerator until cold.

3 Add a few ice cubes to the soup just before serving. Serve accompanied by the garnishes, each in a separate dish.

Cook's tips

This iced soup used to be prepared by hand, using a pestle and mortar. The advent of blenders and food processors has changed all that and now gazpacho *is quick and easy to make. The vegetables and combination of herbs used may be varied according to taste and also to what is in season.*

128

Paella valenciana

- **Preparation: 1 hour**
- **Cooking: 1 hour**

200ml/7fl oz olive oil
1.5kg/3¼lb chicken, cut into 8
    pieces
1 onion, chopped
2 garlic cloves, peeled and chopped
1 red pepper, seeded and cut into
    strips
4 small tomatoes, skinned, seeded
    and chopped
1-2 squid, backbones and heads
    removed
400g/14oz medium-grain Spanish
    or Italian rice
1L/1¾pt chicken stock
2 pinches of saffron threads
8 raw prawns
salt and pepper
100g/4oz frozen peas
8 clams or mussels in their shells,
    washed and scrubbed
lemon wedges, to garnish

- **Serves 4**
- **1150cals/4830kjs per serving**

GOLDEN WONDER

Saffron is the dried stigma of the saffron crocus. It gives a strong yellow colour and unique flavour to dishes. It is very expensive, but fortunately only a pinch is needed. The threads are soaked in a little liquid for 15 minutes or lightly toasted in a dry pan before using.

Avoid powdered saffron as the flavour is not so good.

1 Heat the oil in a large frying pan over medium heat and, when hot, add the chicken pieces. Fry them until golden brown on all sides, about 8-10 minutes. Remove the chicken from the pan with a slotted spoon and reserve it.

2 Add the onion, garlic, red pepper and tomatoes and fry, stirring occasionally, while you prepare the squid.

3 Cut the squid into rings and add these, with the tentacles, to the oil in the pan. Cook gently over low heat for 5 minutes, stirring often. Add the rice, spread it over the contents of the pan and cook for 1 minute, stirring.

4 Meanwhile, bring the stock to the boil. Toast the saffron threads carefully in a small dry pan, then crush and add them to the stock. Pour the stock into the frying pan and stir well. Stir in the unpeeled prawns and bring the mixture to the boil. Season with salt and pepper to taste, add the chicken and cook over low heat for 15 minutes without stirring.

5 Stir in the peas and add the clams or mussels to the pan, pressing them into the rice. Cover the pan and cook for 5 minutes. The shellfish will cook in the steam and should open (discard any that do not).

6 Continue cooking the paella uncovered for 10 minutes or until all the liquid has been absorbed and the surface is quite dry. Turn off the heat, cover the pan with a cloth and leave the contents to settle for 5-10 minutes. Serve from the pan, garnished with lemon wedges.

Variations

Ingredients for a paella can vary – the only basics are rice, twice the volume of liquid, olive oil and saffron. Other paella ingredients to use might include lobster, rabbit, eels, chorizo, snails, artichokes or green beans.

Cook's tips

In a domestic kitchen it is not practical to make paella for more than six people as you will need a very large gas ring to give the right amount of spread-out heat to cook the paella evenly.

Sea bass with orange sauce

- *Preparation: 25 minutes*

- *Cooking: 45 minutes*

*4 × 100-175g/4-6oz skinned and
boned sea bass steaks, skin and
bones reserved, cut from tail end
of fish*
1 bay leaf
salt and pepper
3tbls olive oil
*orange slices and flat-leaved
parsley, to garnish*
For the sauce:
50g/2oz butter
25g/1oz flour
*150ml/¼pt freshly squeezed
orange juice (1-2 oranges)*
juice of 1 small lemon
grated zest of 1 orange

- *Serves 4*

- *365cals/1535kjs per serving*

1 Simmer the skin and bones of the sea bass in 200ml/7fl oz water with the bay leaf for 20 minutes. Strain, season and reserve the stock.

2 Heat the oil in a frying pan over medium heat and fry the fish for 5 minutes each side or until cooked. Drain the steaks on absorbent paper, transfer them to a hot serving dish and keep warm.

3 To make the sauce, melt the butter in a small saucepan over low heat and stir in the flour. Gradually add 150ml/¼pt of the fish stock and the orange and lemon juice, stirring all the time to avoid lumps. Stir in the orange zest, season to taste and stir continuously over low heat until the sauce is smooth. Pour the sauce over the fish steaks and garnish.

Cook's tips

You can substitute the juice from Seville oranges, when in season, for the lemon and orange juice.

SWEET SOMETHINGS

It is generally the custom in Spain to eat fresh fruit such as figs, oranges or grapes to end a meal. The sweetmeats made of almonds, sugar and eggs which are so popular in the south are reserved for teatime or special occasions.

FEELING SAUCY

Sometimes Madrid stew is served with a very simple tomato sauce: fry 1 very finely chopped onion with 2 crushed garlic cloves and 25g/1oz finely chopped parsley in 50ml/2fl oz olive oil until the onion is soft. Add 400g/14oz canned tomatoes, ½tsp crushed cumin seeds and a large pinch of sugar, salt and pepper. Stir, add 1tbls wine vinegar, cover and simmer for 10 minutes. Strain and simmer until thick. This keeps well for up to 2 weeks stored in the refrigerator.

Madrid stew

This is one of Spain's oldest national dishes. It is very satisfying and rather rich and is a whole meal in itself. The broth is traditionally served separately before the meat and vegetables

- *Preparation: 30 minutes, plus overnight soaking*
- *Cooking: 2¾ hours*

225g/8oz chickpeas, soaked
  overnight
½ boiling chicken
450g/1lb lean brisket of beef in one
  piece
225g/8oz lean smoked bacon
  knuckle
1 bay leaf
100g/4oz morcilla or other blood
  sausage
225g/8oz chorizo sausage
½ bulb of garlic, unpeeled, cloves
  unseparated

1 large onion, roughly chopped
½ small cabbage, coarsely
  shredded
2 carrots, quartered
the whites of 2 leeks with some of
  the green part, chopped
2 large potatoes, in chunks
salt and pepper

- *Serves 4-6*
- *850cals/3570kjs per serving*

1 Rinse the chickpeas and tie them loosely in a muslin bag. Put them into a very large pot with the chicken, beef, bacon and the bay leaf. Cover with water, bring to the boil, skim, then simmer, partially covered, for 1½ hours.

2 Add the sausages and garlic head and cook gently for 30 minutes, keeping the pot topped up with boiling water all the time.

3 Add the onion, cabbage, carrots, leeks and potatoes and continue to simmer for about 30 minutes or until the vegetables, chickpeas and meat are all tender. Add pepper to taste and, if wished salt, but be sparing as the bacon might be quite salty anyway.

4 To serve, discard the bay leaf, the garlic head and the knuckle bone. Carefully transfer the chicken, the beef and the sausages to a carving board and cut them into neat slices. Arrange them on a heated serving dish. Remove the chickpeas from the muslin bag and pour them in

a mound in the centre of another large heated dish. Surround the chickpeas with the rest of the vegetables and then serve immediately.

Cook's tips

Tender pre-cooked pieces of pasta can be served in the broth as a separate course.

Wrinkled potatoes with garlic sauce

- *Preparation: 30 minutes*
- *Cooking: 30 minutes*

700g/1½lb walnut-sized new
  potatoes
25g/1oz coarse salt
For the sauce:
4 large garlic cloves
1tsp cumin seed
1tsp paprika
pinch of dried thyme
150ml/¼pt olive oil
2tsp vinegar

- *Serves 4*
- *440cals/1850kjs per serving*

1 Make the sauce first as it has to cool thoroughly before use. Crush the garlic to a pulp with a pestle and mortar. Add the cumin seed and crush it well. Add the paprika and thyme and mix well. Transfer to a small bowl.

2 Slowly add the oil to the garlic, drop by drop at first, as for mayonnaise, stirring all the time and adding the vinegar at intervals. When all the oil and vinegar is used up, whisk in about 50ml/2fl oz warm water to make a thin sauce. Stir well, pour into a serving bowl, cover and set aside until required. 🕐

3 Wash the potatoes thoroughly but leave the skins on. Put them on to boil with just enough water to cover them, leaving the lid off, so that the water evaporates.

4 When the water has almost gone, (this will take about 20-25 minutes), throw in the salt, which will form a crust on the skins. Let the water boil away over low heat, watching it carefully. When it has all evaporated, leave the heat on for about 1-2 minutes or so to dry the potatoes thoroughly and to get a wrinkled effect. Shake the pan a little if they threaten to burn.

5 Serve the potatoes in individual shallow earthenware dishes. Everyone spears his or her own potatoes with a fork and then dips them into the garlic-flavoured sauce.

Serving ideas

Serve this dish as an accompaniment to plainly cooked meat, fish or poultry. The sauce is also delicious trickled over a piece of crusty bread.

Valencia chocolate mousse

- **Preparation: 25 minutes**
- **Cooking: 20 minutes, plus chilling**

4 eggs, separated
4tbls milk
50g/2oz caster sugar
15g/¹/₂oz ground almonds
15g/¹/₂oz flour
50g/2oz butter, cut into small pieces
juice and grated zest of 1 orange
¹/₂tsp powdered gelatine
200g/7oz plain chocolate, broken in pieces
For the decoration:
150ml/¹/₄pt double cream, whipped
2 slices of orange, quartered

- **Serves 4**   🍴 ££ 🕐
- **750cals/3150kjs per serving**

CHURROS CHOICE

Sweet deep-fried fritters known as *churros* are a very popular snack in Spain. The dough is cooked in a sausage-shaped spiral which is then cut up into pieces and liberally sprinkled with sugar. It is sometimes eaten at breakfast time served with steaming mugs of luxurious hot chocolate. This is made with melted pieces of plain dark chocolate and milk.

1 Blend the egg yolks and milk together in the top pan of a double boiler or a bowl set over a pan of hot water. Stir in the sugar, ground almonds and flour and heat very gently, stirring all the time, until the mixture thickens.

2 Add the pieces of butter, one at a time, stirring well as you go. Add the orange juice and zest and sprinkle on the gelatine. Add the chocolate piece by piece, stirring until it is melted and everything is smoothly blended together. Remove from the heat.

3 In a large, clean, dry bowl, whisk the egg whites until stiff. Pour the chocolate mixture into a large bowl and fold in the egg whites. Turn the mixture into four dessert glasses or ramekins and chill for several hours. 🕐

4 Before serving, decorate the mousses with the whipped cream and the pieces of orange.

Almond biscuits

These soft little biscuits, known as *tejas de almendras,* are traditionally eaten at Easter time.

- **Preparation: 30 minutes**
- **Cooking: 20 minutes**

150g/5oz caster sugar
150g/5oz butter, at room temperature, plus extra for greasing
50g/2oz almonds, blanched, lightly toasted and chopped
100g/4oz flour, sifted
25g/1oz ground almonds
pinch of salt
grated zest of 1 lemon
4 egg whites

- **Makes about 16**   🍴 ££ 🕐 ❄
- **160cals/670kjs per biscuit**

1 Heat the oven to 180C/350F/gas 4. Cream the sugar and butter together until light and fluffy. Stir in the chopped almonds, flour, ground almonds, salt and lemon zest.

2 In a large, clean, dry bowl, lightly whisk the egg whites. Fold them into the almond mixture. Drop teaspoons of the mixture onto well-buttered baking sheets, spacing them out well. Bake for about 15-20 minutes until golden, then transfer to a wire rack to cool. 🕐

Pyrenean Plenty

This is the land of d'Artagnan and Armagnac, the mysterious Basques and France's best-loved king, Henri IV, who wanted every Frenchman to have a chicken in the pot on Sundays

*T*HE PYRENEES STRETCH right across this part of France, linking the Mediterranean to the Atlantic and dividing the Basques between France and Spain. Gascony and Foix are two of the oldest French provinces, rich in history and tradition, and with an accent that is easily recognizable throughout France.

Basque separatism
On both sides of the mountains, the Basques cling to their unusual language and customs and distinctive cuisine. The fertile farms produce the delicious tomatoes and peppers which form the basis of the sauce traditionally accompanying dishes known as *basquaise* – the Basque omelette, Piperade, and Chicken basquaise (also with ham and wine).

For generations, the Basques have been great sailors and in early times were famous for hunting whales in open boats. Nowadays they restirct themselves to local tuna and sardines, but fish dishes are a great speciality and salt cod prepared

The classic French chicken in the pot

with tomatoes, pimentos and masses of garlic is a classic dish on both sides of the Pyrenees (page 136).

Gascon panache
Gascony was once a prized English possession, part of the dowry of Eleanor of Aquitaine when she married Henry II, but it has had little influence on English cooking! Up on the high mountain peaks, there is still wild game for the hunting, but

modern cooks will usually prefer farmyard food – ducks are a speciality, raised for their livers but with plump breasts which are delicious cooked in a piquant Gascon sauce (opposite). The cold mountain winters encourage the preservation of meats in the form of cured hams, country sausages and pâtés. Some of these delicious dishes will be enriched with local truffles.

The best of Béarn

This is a land which favours hearty trenchermen and one of the typical country dishes of the entire area is the soup known as Garbure (gammon and vegetable soup, opposite) which is traditionally so thick that a spoon will stand up in it. Here too, poultry and pork are preserved – Bayonne ham, cured in salt, easily rivals Spanish serrano ham, and both geese and ducks end up as *confit* in great earthenware jars, waiting to flavour stews throughout the year. For most people, though, it is sauce Béarnaise (made with shallots, egg, butter and herbs) which reminds them of the local cooking. This warm sauce is a perfect accompaniment.

Chicken in the pot

Henri IV wished every family could afford this dish on Sunday.

● **Preparation: 1 hour**

● **Cooking: 3 hours including cooking stock**

2.5kg/4¹/₂lb boiling fowl with giblets
1 beef marrow bone
1 large onion, peeled
1 stalk celery
1 garlic clove
1 bouquet garni (3 parsley sprigs, 1 sprig rosemary and thyme, and 1 bay leaf)
225g/8oz carrots, coarsely sliced
225g/8oz turnips, coarsely sliced
white part of 1 leek, coarsely sliced
6 outer leaves from 1 large green cabbage
6 medium-sized potatoes, peeled
salt and freshly ground black pepper

Preparing the famous salt-cured jambon de Bayonne in the traditional way

For the stuffing
1 chicken liver from the boiling fowl
25g/1oz butter
75g/3oz crumbled, stale bread
2tbls milk
225g/8oz smoked gammon, diced
2 garlic cloves, crushed
1 shallot, finely chopped
25g/1oz parsley, chopped
pinch of dried tarragon
pinch of dried chervil
pinch of ground nutmeg
2 medium-sized eggs, beaten
salt and freshly ground black pepper

● **Serves 6 for soup and main course**
🍴 ££

● **635 cals/2665 kjs per serving**

1 First prepare the stock: put all the giblets (except for the liver) into a casserole or saucepan large enough to hold the chicken, with the marrow bone, onion, celery, garlic and bouquet garni. Add 3L/5pt water to the pan, cover and simmer for at least 1 hour.

2 For the stuffing, cook the trimmed liver in the butter in a small saucepan over low heat until browned on all sides but slightly pink inside, about 3 minutes. Soak the crumbled bread in the milk, chop the

liver and mix it with its cooking liquid, the bread, gammon, garlic, shallot, parsley, tarragon, chervil and nutmeg. Bind the mixture with the eggs and season with salt and freshly ground pepper. Reserve 60g/2½oz of the stuffing and put the rest inside the bird and truss it firmly.

3 Strain the stock, return it to the saucepan, bring it to the boil over high heat and lower the chicken into it. When the stock comes to the boil again, add the coarsely sliced carrots, turnips and the ▶

134

white of the leek, cover and simmer for about 1½ hours.

4 Meanwhile, place the cabbage leaves in a large bowl, pour boiling water over them and leave them in the water until soft, about 5 minutes. Drain the leaves and allow them to cool. Place 1tbls of the reserved stuffing in the centre of each cabbage leaf, roll them into parcels and secure them with cotton thread or wooden cocktail sticks.

5 When the chicken has cooked 1½ hours, add the potatoes with the cabbage parcels to the chicken and cook until the potatoes are tender, about 20-30 minutes. Transfer the chicken and vegetables to a platter and keep warm. Check the seasonings of the broth, then serve it separately for the first course.

Duck breasts in piquant sauce

A luxurious dish with an interesting piquant flavour, in Gascony it would be made with the breasts of ducks specially fattened for their livers. However, plump supermarket duck breasts do very well.

- ● *Preparation: 15 minutes*

- ● *Cooking: 25 minutes*

  2 plump duck breasts, boned with the excess fat removed and reserved and the skin well pricked

75ml/3fl oz Armagnac
40g/1½oz flour
1tsp tomato purée
150ml/5fl oz white wine
100ml/3½fl oz chicken stock
salt
100ml/3½fl oz double cream
20 preserved green peppercorns, drained

- ● *Serves 2*

- ● *735 cals/3085 kjs per serving*

1 In a small saucepan, render the reserved duck fat to produce 2tbls liquid fat. If using ready-prepared duck breast, use butter.

2 Transfer the fat to a heavy frying-pan and brown the breasts over medium heat, skin side down first, for 2-3 minutes each side. Pour off the fat (the skin will make at least 1tbls extra fat) and reserve.

3 Flame the breasts with the Armagnac and transfer them to an ovenproof dish at the bottom of the oven. Heat the oven to 180C/350F/gas 4.

4 Return 3tbls of the reserved duck fat to the pan over medium heat. When sizzling, stir in the flour and allow it to colour well. Stir in the tomato purée, the wine and stock and season to taste with salt. Simmer the mixture for 5 minutes or until slightly thickened.

5 Pour the sauce over the duck breasts, cover the dish with foil, transfer it to a shelf just above the centre of the oven and

cook for 25 minutes.

6 Remove the dish from the oven, stir in the cream and green peppercorns to the sauce and check the seasoning. Return the dish, uncovered, to the oven for a further 5-10 minutes. Serve immediately.

Gammon and vegetable soup

Typical of the Gascony Pyrenees, this soup makes a meal in itself. The vegetables can be varied according to what is in season: the only constant factor is the shredded cabbage – either green or white.

- ● *Preparation: 30 minutes*

- ● *Cooking: 1 hour*

450g/1lb unsmoked gammon, in 1 piece
4 garlic cloves, peeled
2 large potatoes, peeled and thickly sliced
4 carrots, washed
2 onions, thickly sliced
450g/1lb green beans
250g/9oz broad beans
sprig of thyme
sprig of parsley
salt and freshly ground black pepper
250g/9oz cabbage, finely shredded
4 thick slices stale brown bread

- ● *Serves 4*

- ● *455 cals/1910 kjs per serving*

1 Put the gammon in a large saucepan, pour in enough cold water to cover the gammon and bring up to the boil.

2 When boiling briskly, add the garlic, potatoes, carrots, onions, beans and herbs. Bring to the boil again, cover and simmer for 30-40 minutes.

3 Add salt and freshly ground black pepper to taste, remove the herbs and add the cabbage. Bring to the boil, cover the pan and simmer for a further 10-20 minutes.

4 When ready to serve, put a slice of bread at the bottom of each of 4 individual warmed soup bowls. Remove the gammon from the saucepan, cut it into thick slices and divide equally amongst the bowls. Place the thick slices of gammon on top of the bread.

5 Spoon the vegetables and the broth evenly over the bread and gammon and serve piping hot.

Salt cod, Biscay-style

This classic Basque dish, equally popular on both sides of the Franco-Spanish frontier, is traditionally made with the dried and cured cod that is a staple not only of Basque cooking but that of all Spain and Portugal. It can be found in delicatessens and must be soaked in cold water overnight before using. Thick fillets of smoked cod can be substituted but if they are too heavily smoked, they can give a very fishy taste.

● **Preparation: 15 minutes, plus overnight soaking**

● **Cooking: 25 minutes**

1kg/2¼lb dried salt cod, soaked in cold water overnight, or smoked cod
450g/1lb tomatoes, fresh or canned
5 garlic cloves
150ml/5fl oz olive oil
450g/1lb onions, chopped
½tsp ground hot red chilli pepper
salt and freshly ground black pepper
a little flour
25g/1oz stale bread, grated into crumbs
25g/1oz parsley, chopped
200g/7oz canned pimentoes

● **Serves 6**

● **375 cals/1575 kjs per serving**

1 Blanch and skin the fresh tomatoes, if using, removing all seeds and juice; chop the flesh. Drain canned tomatoes and chop. Crush the garlic cloves with the side of a knife and put the garlic and tomatoes in 100ml/3½fl oz of oil in the saucepan, together with onions, chilli pepper and a very little salt and freshly ground black pepper. Place the saucepan over low heat and simmer until the mixture is very soft.

2 Meanwhile, heat the oven to 190C/375F/gas 5 and drain the salt cod. Put the fish in a saucepan and cover with cold water. Over medium heat bring to the boil, then barely simmer for 2-3 minutes.

3 Remove the cod carefully with a spatula and allow it to cool for 5-10 minutes. Bone and skin the cod and cut it across into fingers 2.5cm/1in wide, discarding any skin. Pat them dry with kitchen paper, flour them lightly and reserve.

4 When the sauce has reduced to the consistency of a thick purée, check the seasoning and sieve half of the sauce into a shallow ovenproof dish. The sauce can be liquidised or put through a food processor and then sieved or not, as liked. Arrange the fish portions on the sauce and sieve the remaining sauce on top.

5 Fry the breadcrumbs lightly in the remaining oil and sprinkle with the parsley over the top layer of sauce.

6 Drain the pimentoes and slice them into thin strips. Arrange a pattern on top of the parsley and breadcrumbs and place the dish in the oven 5-7 minutes or just long enough for the surface to toast without it becoming too brown.

Basque stuffed tomatoes

This is a striking-looking vegetable accompaniment to roast or grilled meat; use Mediterranean-style tomatoes, as big as a small fist.

- **Preparation: 25 minutes**

- **Cooking: 25 minutes**

4 large tomatoes
125g/4½oz slightly salted butter
450g/1lb onions, finely chopped
4 green peppers, seeded
salt and freshly ground black
   pepper
25g/1oz fresh breadcrumbs

- **Serves 4**   (🍴)(££)

- **300 cals/1260 kjs per serving**

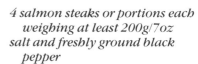

1 Cut the tomato tops off evenly and scoop out the cores, pips and juice with a sharp-pointed teaspoon. Turn the tomatoes upside-down on absorbent paper and allow to drain.

2 Meanwhile, melt 50g/2oz butter in a saucepan over low heat and cook the onions until soft. Heat the grill to high.

3 Cut the peppers in half lengthways and flatten them on the grill pan, skin upwards. Grill them quickly for a few minutes until the skin starts to char and blister. Remove from heat and when they are cool enough to handle, peel and cut them into thin slices. Put 50g/2oz butter in another saucepan with the pepper slices, cover and place over low heat to soften, about 10 minutes.

4 Heat the oven to 140C/275F/gas 1. Lightly salt and pepper the insides of the tomatoes and two-thirds fill them with the onions. Lightly salt and pepper the onion layer and cover it with pepper strips which should rise in domes above the tops of the tomatoes. Sprinkle the tomatoes with breadcrumbs, top with thin slices of the remaining butter. Place the tomatoes on a baking tray and put in the oven for 15-20 minutes.

Grilled salmon steaks Béarnaise

This is a spectacular dish for a summer lunch party, when the salmon can be grilled in the traditional way over charcoal while the sauce (made in advance indoors), is kept tepid over hot water.

- **Preparation: 15 minutes**

- **Cooking: 45 minutes**

4 salmon steaks or portions each
   weighing at least 200g/7oz
salt and freshly ground black
   pepper
olive oil
For the Béarnaise sauce
2 shallots, finely chopped
2 parsley sprigs
75ml/3fl oz dry white wine
50ml/2fl oz tarragon vinegar
salt and freshly ground black
   pepper
3 medium-sized egg yolks
150g/5oz butter, melted
4-5 tarragon leaves, finely chopped,
   or ¼tsp dried tarragon
pinch of dried chervil
pinch of cayenne pepper
a few drops of lemon juice

- **Serves 4**   (🍴)(£££)

- **715 cals/3005 kjs per serving**

1 Start the sauce first: put the shallots, parsley, wine, vinegar and a grating of black pepper into a small saucepan over medium heat and simmer until reduced by two-thirds to about 3tbls. Strain the liquid into a bowl that fits over a saucepan and let it cool slightly.

2 Meanwhile, lightly whisk the egg yolks with 1tsp cold water. Mix them gradually into the tepid wine mixture and place the bowl over a saucepan of barely simmering water over very low heat; the water in the pan must not touch the bottom of the bowl.

3 Whisk the egg yolk mixture until it starts to thicken and slowly add the ▶

melted butter, whisking continuously. When the mixture becomes thick and creamy (this happens very quickly), turn off the heat and add a little salt to taste. Keep the sauce warm on the pan.

4 Season the salmon with salt and freshly ground black pepper. Grill the salmon for about 5 minutes, brushing lightly with oil on each side during cooking to prevent them from drying. Carefully remove the skins.

5 Just before serving the sauce, stir in the tarragon and chervil with the cayenne pepper dissolved in a few drops lemon juice. Pour over the fish steaks. Garnish with parsley if wished and accompany with boiled potatoes and a tossed green or mixed salad.

BEANS MEANS ICE CREAM

Although Gascony and Foix are not short of fruits for their ice cream, one of their local specialities is made from haricot beans. They are cooked with demerara sugar, puréed and blended with creamy custard and the local Armagnac. Sounds unusual ... but apparently it has a delicious nutty flavour.

Basque cherry tart

Traditional gâteau basque is a tart of unusual, cake-like pastry filled with crème pâtissière and the exquisitely flavoured cherries of the Itxassou region. Substitute 700g/1½lb canned Morello cherries if you wish, removing the stones and drying the fruit well with absorbent paper.

● *Preparation: 15 minutes, plus 2 hours resting time*

● *Cooking: 45 minutes*

175g/6oz self-raising flour
2 small eggs, separated
100g/4oz butter, diced small
100g/4oz caster sugar
pinch of grated lemon zest
pinch of salt
450g/1lb cherries, stoned
1tbls redcurrant jelly

● *Serves 4*

● *530cals/2225kjs per serving*

1 Sift the flour into a bowl and make a well in the centre. Add the egg yolks plus 3tsp egg white, beaten together, with the butter, sugar, grated lemon zest and

salt to the well in the flour.

2 Work the ingredients in the centre together with a fork, gradually taking in the flour, then mix with the fingertips until the dough is fairly firm or use a food processor. Cover the dough with a cloth and chill for at least 2 hours (longer if convenient).

3 Heat the oven to 170C/325F/gas 3. Divide the pastry in half. Roll out one half and line a 18cm/7in flan tin, letting the pastry edges overlap the tin. Fill the tin with the fruit and dot the jelly on top. Roll out the remaining pastry and cover the tart. Fold the overlapping pastry inwards and crimp the edges.

4 Pierce a few holes in the top of the pastry with a fork and glaze the top with the rest of the egg white, lightly beaten. Bake the tart in the oven for 30-40 minutes or until lightly golden.

CHERRY RIPE

Local fruit throughout the whole area is notable for its variety and flavour – especially delicious are the sharp cherries of Itxassou which make the Basque cherry tart (given on this page) an unusual experience. In season, the markets are full of fragrant apricots, plums and quinces. An unusual product is angelica, which grows wild in the Pyrenees and is preserved in sugar for use in confectionery.

Swiss Selection

Swiss dairy cows graze on lush grass in Alpine meadows and produce a high-quality milk which goes to make some of the most wonderful cheeses and chocolate in the world

*C*OMPLETELY SURROUNDED BY other countries, Switzerland is naturally greatly influenced by its neighbours: it has three different main languages and cultures. This influence is no less strong when it comes to the cuisine: in the west the cooking is influenced by France, in the south by Italy and in the north by Germany. Switzerland does have its own heritage and range of national dishes, however, most of them using cheese.

There are also many regional dishes based on local products such as lake fish or fruit. Recipes for these traditional specialities and local dishes have been passed down from generation to generation and they are still made by city dwellers who have all the modern conveniences and a wider choice of ingredients than the original cooks ever dreamed of.

Veal is probably the most popular meat in all areas but, over the winter months, the wide range of game birds and venison from the Grisons in eastern Switzerland adds variety to the menu.

Cheese fondue (page 142)

There is also the celebrated *Bündnerfleisch,* or air-dried meat, which is a speciality of several districts. It is served in paper-thin slices as an appetizer and it is not easy to find outside of Switzerland.

Switzerland's lush pastures provide grazing for great herds of dairy cattle which produce the rich milk used in famous Swiss cheeses. The history of Swiss cheese goes back to the time when Caesar's legions took great wheels of cheese back home

after campaigns. The early cheeses were made with a heavy rind which thickened with age and became almost like leather. This rind protected the cheese against weather and time, so the cheese could be eaten by mountain farmers the following winter.

Time-old traditions

Many traditions, based on particular cheeses, evolved in mountain villages and these have been handed down for centuries. One such tradition involves a farmer setting aside a wheel of cheese when his son is born. The cheese may only be cut for his christening, wedding and funeral.

Probably the best known of Swiss cheeses is Emmental, a firm, pale yellow cheese with a nutty taste and characteristic large holes. Gruyère, often mistaken for Emmental, has smaller holes and a brown crinkled rind; it is creamier and rather more acid tasting. It originally came from the Gruyère valley in the canton of Fribourg. Both cheeses cook well and combine to make the most famous Swiss dish, Cheese fondue (see recipe).

The word fondue comes from the French *fondre* meaning to melt, but it is not known which of the three French-speaking cantons first made it, using their local cheeses Emmental, Gruyère and Vacherin. The Swiss maintain that a real fondue is made with one or a mixture of these cheeses, plus kirsch – the potent Swiss cherry *eau de vie* – and a white wine. However, other cheeses

Applying the name to a wheel of Emmental cheese

can be used with gin or vodka instead of kirsch; sometimes dry white wine alone is used.

Fondue fun

A fondue is an ideal party dish and is almost a ceremony with its own rules. It is made in a special communal pot known as a *caquelon* which is made of either earthenware or enamelled iron with a rounded inside base and a short handle.

Each person spears a cube of bread on a special long-handled fondue fork and dips it into the hot fondue. The bread is twirled around to catch the drips and transferred to a plate to eat with an ordinary fork. This is more hygienic and also saves a burnt mouth!

Raclette is another popular cheese dish in Switzerland, especially for informal winter parties. A round of cheese that melts quickly, usually from the cantons of Conches or Bagnes, is cut into two and the cut side held in front of an open fire or a special charcoal grill. When the cheese melts, it is quickly scraped off onto a warmed plate, then eaten with boiled new potatoes, pickled onions and a Swiss white wine.

The fertile Swiss valleys grow prime vegetables and fruit, while vineyards on the sun-facing slopes, notably in the French-speaking regions of Vaud, Valais and Neuchâtel, produce fine wines. The most popular wines are the white Fendant and the red Dôle.

Swiss chopped veal

- ● *Preparation: 25 minutes*
- ● *Cooking: 15 minutes*

450g/1lb veal escalopes
flour, for coating
salt and pepper
50g/2oz butter
1 large onion, finely chopped
2tbls lemon juice or white wine

- ● *Serves 4*
- ● *210cals/880kjs per serving*

1 Cut the meat into 5mm/¼in cubes, sprinkle lightly with flour, salt and pepper and reserve.

2 Melt the butter in a frying pan over medium heat, then sauté the onion until it is tender but only lightly browned.

3 Add the meat to the pan, raise the heat to medium-high and cook fairly quickly for 3-4 minutes, turning frequently. The small pieces of meat should cook in that time. Stir in lemon juice or wine and serve at once, with Rösti (see recipe).

Rösti

This dish of fried potatoes is served all over Switzerland, sometimes as an accompaniment to meat, sometimes as a meatless meal. Of its many variations, here is a favourite one of the Swiss German-speaking area

- *Preparation: 20 minutes*
- *Cooking: 45 minutes*

6 floury potatoes
1 onion, finely chopped
3-4 slices lean bacon, finely chopped
50g/2oz butter
salt and pepper

- *Serves 4-6*   ① ⑤
- *290cals/1220kjs per serving*

1 Boil the potatoes in their skins for 20 minutes or until they are just tender.

2 Remove the pan from the heat, drain the potatoes in a colander and cool them under running cold water. When the potatoes are cool enough to handle, strip off the skins.

3 Grate the skinned potatoes into a large bowl, using the coarsest blade of the grater. Then carefully mix in the chopped onion and bacon, salt and pepper and reserve.

SLOWLY BUT SURELY

Unless you go to Switzerland you are unlikely to see many Swiss cheeses other than Emmental and Gruyère, but the Swiss actually make more than 100 different varieties of cheese. They are still produced using old-fashioned methods by family-owned dairies, so one dairy might only produce two wheels of Emmental a day.

4 Heat the butter or lard in a large, heavy frying pan about 23cm/9in across. When the fat is quite hot, add the potato mixture. Lightly pat the mixture into a round cake with a spatula or fish slice. Cook over low heat for 10-15 minutes or until the underside is crusty and browned.

5 Turn the potato mixture onto a warmed plate and flip it back into the frying pan. Fry gently for 10-15 minutes until the second side is crisp and well browned. Turn the Rösti out onto a heated serving dish, cut into wedges and serve.

Pea-pod soup

- *Preparation: 30 minutes*
- *Cooking: 35 minutes*

15g/½oz butter
2tbls chopped onion
450g/1lb pea pods, topped and tailed
3-4 large lettuce leaves
850ml/1½pt chicken stock
150ml/¼pt single cream
salt and pepper
croûtons or finely snipped chives, to serve

- *Serves 4-6*   ① ⑤ ⏱ ❋
- *305cals/1280kjs per serving*

1 Melt the butter in a large saucepan over very low heat, add the chopped onion, cover and cook gently for 8-10 minutes, until cooked but not brown.

2 Add the pea pods, lettuce leaves and stock to the pan and bring the mixture to the boil. Reduce the heat and simmer, covered, for 20 minutes or until tender. Purée the mixture in a blender or pass it through a sieve. ⏱

3 To serve the soup hot, return it to the saucepan, season with salt and pepper, add the cream and reheat it without boiling. Serve in warmed bowls with croûtons.

4 To serve cold, chill the soup in the refrigerator for at least 2 hours, then ▶

◀ skim off any fat. Divide the soup between individual bowls, swirl a little cream into the top of each bowl, sprinkle with snipped chives and serve.

Chocolate cream cake

- **Preparation: 45 minutes, plus overnight chilling**
- **Cooking: 5 minutes**

100g/4oz plain chocolate, broken into pieces
175ml/6fl oz milk
75g/3oz butter
100g/4oz caster sugar
1 egg yolk
300ml/¹⁄₂pt whipping cream
24 sponge fingers
For the decoration:
150ml/¹⁄₄pt double cream, whipped
grated chocolate

- **Serves 6-8**
- **505cals/2120kjs per serving**

1 Line a 1.5L/2½pt loaf tin with foil or thick greaseproof paper, leaving a good border each side to lift out the cake.

2 Melt the chocolate in half the milk in a saucepan over low heat, removing it as soon as the chocolate is melted. Stir in the remainder of the milk and reserve.

3 Cream the butter and sugar together until they are light and creamy, beat in the egg yolk and then the chocolate mixture, and continue beating until the

Pea-pod soup (page 141)

mixture is smooth. Whip the cream until it is thick.

4 Pour one-third of the chocolate mixture into the lined tin, cover with a layer of sponge fingers, then a third of the whipped cream. Repeat with the rest of the layers, finishing with whipped cream. Chill overnight.

5 When ready to serve, turn the cake out onto an oblong serving plate, cover the sides with two-thirds of the remaining whipped cream and decorate with grated chocolate. If you wish, use the remaining cream to pipe a border around the edge of the cake.

CHOCOLATE FONDUE

Coarsely grate milk chocolate with nuts, preferably Swiss. Melt it in a little milk or single cream, then stir until quite smooth. Cubes of sponge cake, peeled and cubed bananas, apples, pears or pieces of pineapple are all delicious dipped in the hot chocolate. Be careful of burned mouths!

Cheese fondue

- **Preparation: 30 minutes**
- **Cooking: 15 minutes**

1 garlic clove
350g/12oz Gruyère cheese
350g/12oz Emmental cheese
2tsp cornflour
2tbls kirsch
425ml/³⁄₄pt dry white wine, more if needed
2tsp lemon juice
pinch of white pepper
pinch of grated nutmeg
1 large loaf of crusty French bread, cut into 2.5cm/1in cubes, for dipping

- **Serves 4-6**
- **1165cals/4895kjs per serving**

1 Cut the garlic clove in half and use it to rub round the inside of the fondue pot. Grate the cheeses fairly coarsely. Blend the cornflour with the kirsch until smooth; reserve.

2 Over high heat on the kitchen hob, heat the wine and the lemon juice in the pot until just boiling, then turn the heat to low and stir in the cheese very slowly with a wooden spoon. Add the kirsch mixture, pepper and nutmeg, stirring constantly until thick. The cheese mixture should be smooth; stir in a little more warmed wine if necessary.

3 Transfer the pot to a spirit stove or electric hot plate at the table and keep it simmering; do not allow it to boil. Each person can now dip pieces of French bread into the bubbling cheese, using long-handled fondue forks. Remember to stir the pot frequently during the meal.

4 When nearly all the fondue is eaten there will be a thick crust on the bottom of the pot: this should be scraped out and divided between the guests.

INDEX

ACKNOWLEDGEMENTS

The publishers extend their thanks to the following agencies, companies and individuals who have kindly provided illustrative material for this book. The alphabetical name of the supplier is followed by the page and position of the picture/s.
Abbreviations: b = bottom, l = left, r = right, t = top.

Susan Griggs: 100. Robert Harding: 16b; Hutchison Library: 122b. Image Bank: 32t, 44tl, 52b, 64t, 74, 80. MC Picture Library: Bryce Atwell 60/61b, 61tr, 66t, 107b; Aughes-Gilbey 134t; Paul Bussell 10b, 12t, 12/13b, 14r, 15, 16t, 18/19b, 28/29, 36bl, 44br, 45, 60t, 62, 64b, 76/77t, 81, 88b, 89, 91, 93tl, 93br, 94, 104b, 105, 117b, 118/119b, 128/129t, 130t, 136/137b, 137tr, 138; Alan Duns 50, 104t; Ray Duns 85, 115, 116t, 119t, 120; Laurie Evans 34/35b, 35tr, 56, 73, 75, 95, 114b, 131br; Robert Golden 41t; Paul Grater 57; John Hollingshead 52t, 53t, 54br, 139; James Jackson 9, 38/39b, 49b, 67, 69b, 76bl, 77b, 98b, 121, 122t; Chris Knaggs 18l, 30,51, 108br, 142t; Jess Koppel 82tl, 83/83b Michael Michaels 84, 99, 102/103b, 127, 130/131b, 132 James Murphy 21, 23, 24, 25/25b, 26b; Peter Myers 11, 20, 22b, 37, 41b, 54tl, 58t, 59, 66b, 72, 78br, 79, 80t, 90t, 106b, 123, 124t, 124/125b, 142b; Alan Newnham 86, 109, 111, 112t, 112/113b; Roger Phillips 17t; Paul Webster 17b, 55, 96/97b, 140/141b; Andrew Whituck 47, 48; Paul Williams 26t, 36t, 65t, 68t, 110b, 133. Photographers Library: 88t. Portuguese Tourist Office: 116b. Spectrum: 28cl, 38t, 68b, 96l, 106t, 128bl. Swiss National Tourist Office: 140t. Welsh Tourist Board: 22t. ZEFA: 10t, 58b, 92,110t. All other pictures MC Picture Library.

Index prepared by INDEXING SPECIALISTS, Hove.